Know Your Airplane

Know Your

Airplane!

BY KEITH CONNES

IOWA STATE UNIVERSITY PRESS • AMES

The author is grateful to the many people in all segments of the aviation industry who contributed information that is to be found in these pages. Particular thanks are extended to Dave Ellis, Trevor Linton-Smith, Jim Griswold, Dan Somers, Roy Lopresti, Fen Taylor, and Paul Uitti.

Illustrations by Donald R. Westland

© 1986 Keith Connes. All rights reserved

Composed and printed by the Iowa State University Press, Ames, Iowa 50010.

No part of this book may be reproduced in any form or by any electronic or mechanical means, including information storage and retrieval systems, without permission from the publisher, except for brief passages quoted in a review.

First edition, 1986

ISBN 0-8138-1056-6

Library of Congress Cataloging-in-Publication Data

Connes, Keith, 1926–
   Know your airplane!

   1. Airplanes, Private.   I. Title.
TL685.1.C65   1986   629.133′340422   85-19740
ISBN 0-8138-1056-6

CONTENTS

Acronyms, vi

Introduction, viii

1. Airfoils, 3
2. Flaps, Ailerons, and Spoilers, 13
3. Tails, 20
4. Engines and Fuel, 28
5. Propellers, 41
6. Landing Gear, 47
7. Turbocharging, 58
8. Supplemental Oxygen, 66
9. Pressurization, 76
10. Anti-icing and De-icing Equipment, 85
11. Avionics, 92
12. Instruments, 133
13. Autopilots, 144
14. Mods, 156
15. Airplanes of a Different Sort, 170

Epilogue, 181

Directory of Manufacturers, 183

Index, 189

ACRONYMS

AD	airworthiness directive
ADF	automatic direction finder
AEA	Aircraft Electronics Association
AOPA	Aircraft Owners and Pilots Association
A&P	airframe and powerplant
ASI	airspeed indicator
ATIS	automatic terminal information service
ASTM	American Society for Testing Materials
ATC	air traffic control
CDI	course deviation indicator
CFI	certified flight instructor
CG	center of gravity
CHT	cylinder head temperature
CRT	cathode ray tube
DG	directional gyro
DME	distance measuring equipment
EAROM	electrically alterable read only memory
ECDI	electronic course deviation indicator
EFIS	electronic flight instrument system
EGT	exhaust gas temperature
ETA	estimated time of arrival
ETE	estimated time enroute
FAF	final approach fix
FARs	federal aviation regulations
FBOs	fixed base operators
GMT	Greenwich Mean Time
HSI	horizontal situation indicator
IFR	instrument flight rules
ILS	instrument landing system
KIAS	knots indicated airspeed
LCD	liquid-crystal display
LED	light-emitting diode
LOC	localizer
LORAN	long range navigation system

MDA	minimum descent altitude
MP	manifold pressure
NACA	National Advisory Committee for Aeronautics
NASA	National Aeronautics and Space Administration
NDBs	nondirectional beacons
OAT	outside air temperature
OBS	omni bearing selector
OEM	original equipment manufacturer
PIREPS	pilot reports
PTT	push-to-talk
R & D	research and development
RCC	radio common carriers
RFI	radio frequency interference
RMI	radio magnetic indicator
RNAV	area navigation system
RVP	Reid vapor pressure
SDR	service difficulty report
STC	variously, supplemental type certificate or sensitivity time control
STOL	short takeoff and landing
TACAN	tactical air navigation system
TBO	time between overhauls
TCA	terminal control area
TCP	Tricresyl phosphate fuel additive
TIT	turbine inlet temperature
TSO	technical standard order
TTS	time-to-station
VFR	visual flight rules
VHF	very high frequency
VLF	very low frequency
VNAV	vertical navigation system
VOR	VHF omnidirectional range
VORTAC	combined VOR and TACAN facilities

INTRODUCTION

BACK IN THE GOOD OLD DAYS, when life and airplanes were a lot simpler, a reasonably competent pilot could hop confidently into an unfamiliar light plane with scarcely a glance at the panel, much less at the aircraft's information manual. (The latter wasn't very useful, anyway.)

"It's an airplane, isn't it?" was the standard comment, as our hero reached for the mag switch.

Well, those good old days have been replaced by the better new days, where the airplanes and their equipment are a lot more capable—and a lot more complex.

The information manuals have improved, too. They are organized into a standard format, and they are dedicated to providing you with facts instead of sales pitches. However, in this author's opinion, they don't tell you nearly as much as you need to know in order to properly evaluate and operate the equipment you are commanding.

The major purpose of this book is to help fill that gap, and I've tried to achieve this in a number of different ways.

For example, if you're considering the purchase of, say, a four-place single, you may want to know whether the three-blade prop option is really worth an extra $1,300 and 20 lb. And what about that retractable gear, for $23,000 more than its fixed-gear sister? Is it worth the extra money in performance? How reliable is it, and how effective is the emergency extension system?

Should you run your engine past its TBO (time between overhauls)? What are some simple tips that will keep it healthy longer?

More and more singles are being turbocharged these days. (As of this writing, Piper isn't even making the normally aspirated Arrow.) When does turbocharging really pay off, and when does it just eat up money?

How much laminar flow does a laminar-flow airfoil really produce? Does a T-tail offer any benefits, other than cosmetics? Does it have any drawbacks?

What about those mod shops that claim to give you another 15 kn of cruise, or 200 mi of range, or STOL (short takeoff and landing) performance? If *they* can do it, why don't the aircraft manufacturers do it themselves?

These are some of the questions I have tried to provide authoritative answers for. In the process, I've gone to the working experts: engineers, technical reps, company presidents, and even some marketing types I've learned to trust.

You'll notice that the biggest chapter of this book concerns itself with avionics, and I must tell you it was hard to decide where to draw the line in reporting on the new products. Thanks to microprocessor technology, you have avionics capability at your disposal that was unheard of five years ago. But many pilots don't understand how to get the most out of all those marvelous boxes that are blinking away on the panel. I'll try to provide some input—again, drawing on the aid of many experts.

This book will be useful whether you own or rent an airplane, plan to buy or trade up, or simply like to know as much as you can about the world of general aviation.

A few definitions, disclaimers, and other explanations are in order.

First, this book is directed mostly at pilots or would-be pilots of production aircraft in the single engine and light twin categories. Therefore, I have generally restricted myself to those aircraft and related equipment, although I have brought in some of the heavier iron when I thought it would be of interest.

Most of the aircaft and equipment I've referred to are current models, but I've also included some that have gone out of production in recent years. (Some of those, I'm happy to note, may be coming back!) And I've taken some looks at what the future has for us.

This book is not a directory of all the equipment that's on the market. Rather, the intent is to give you a feel of what is available and how it is used, and to offer some criteria to bear in mind when you're shopping around. With a few exceptions, I have tried to avoid the luxury of substituting my opinions for information, and I hope I've identified my few editorial comments as such.

Realizing that the only consistent thing about prices is that they keep going up (and again, there are a few happy exceptions, particularly when the avionics manufacturers get into their price wars), I have provided price ranges of equipment, rather than item-by-item prices.

With all due respect for women who fly (and I mean a *lot* of respect), I have fallen back on the use of male pronouns. "When he or she decides to get his or her autopilot" just gets too cumbersome. Bear with me, ladies, I am no male chauvinist.

One of the fascinating things about aviation is that if you pay any kind of attention, you keep learning how much you still have to learn. And if you love flying, chances are you love to keep learning about it.

I've been flying for a bit over 28 years, and writing about it a good part of that time. In compiling this book, I have gotten the benefit of a tremendous amount of additional information I know will make me a more competent and more discriminating pilot.

I hope it will be as useful to you.

Know Your Airplane

1. Airfoils

The quest for laminar flow... Reynolds numbers explained ...Which planes have which airfoils...Computers at work.

AN AIRFOIL is a pair of surfaces with a shape that makes your airplane fly. "Well," you might think, "as long as it does its job, why should I want to know anything more about the subject?"

For one thing, your nonflying friends expect you, the pilot, to be knowledgeable about aeronautical matters, and you want to keep the old image polished. But on the more practical side, the various airfoil designs have differing flying and stalling characteristics, and you would be well advised to know which airfoil is used on the plane you fly, and what to expect from it.

Actually, a plane has a number of airfoils, including the wing, stabilizer, control surfaces, and propeller. In this chapter, we'll concentrate on the wing airfoils, and in later chapters we'll explore control surface and propeller designs.

The aircraft designer is faced with the challenge of trying to get a combination of low takeoff and landing speeds, good stall characteristics, good rates of climb, and good cruise speeds. He is also looking for an adequate thickness-to-chord ratio so he can put a substantial spar in the wing.

But the faster the plane is to go, the thinner the wing should be. On the other hand, that will require the wing to be heavier, since the spar must be shaped differently—and fuel volume will be lost to boot. Nor can the designer be totally performance-oriented; there's the inevitable Little Man with the Green Eyeshade, fussing over manufacturing costs.

Soon, we'll see how some designers' dilemmas are resolved.

How an Airfoil Works. Two gentlemen of the 17th and 18th centuries provided a couple of key principles that later helped man get off the ground. The first was Sir Isaac Newton, who pointed out that for every action there is an equal and opposite reaction. Then came Daniel Bernoulli, who discovered that as the velocity of a fluid (such as air) is increased, the static pressure of that fluid is correspondingly decreased.

Now the airfoil comes into play. When a moving airfoil assumes an angle of attack, the result is a nonsymmetrical flow pattern, with the fluid particles moving faster over the top of the airfoil than they are moving over the bottom. Therefore, as per Bernoulli, there is a lower static pressure on the top surface and a positive pressure on the bottom surface—and the result is lift.

An airfoil in the shape of a flat plate can have lift, at certain angles of attack. If the angle of attack is increased beyond a certain point, there will be a large flow separation at the upper surface, resulting in loss of lift in the form of a stall. But a flat plate is far from the ideal airfoil. By adding a rounded leading edge, along with thickness and camber, airflow separation is reduced and the performance of the airfoil is improved at various angles of attack.

As a consequence of the lift on the wing, the airflow is deflected downward, in obedience to Sir Isaac's law. This "downwash" decreases the angle of attack of the horizontal tail and thus influences stability and control.

The Development of Airfoils. In the early days of aviation, airfoil shapes were arrived at pretty much by trial and error—although the Wright Brothers did build their own wind tunnel for testing purposes. Their airfoils were shaped like thin curved plates, in an attempt to duplicate the wings of birds.

Around 1915, Tony Fokker produced an airfoil that had a flat bottom but was curved at the top. The resultant thickness permitted the use of an internal spar, which eliminated much external bracing. Interestingly, in wind tunnel tests made at the time, the Wright Brothers' thin plate airfoil worked better than Fokker's thick airfoil— but these were scale models having wing chords of only 3-4 in., and they didn't give a true picture of what the full-size wings would do. When the wings were built to full scale and flown at higher speeds, it was discovered the curved plates had more drag than the flat-bottom airfoils.

Our government got into the act, most productively. NACA (National Advisory Committee for Aeronautics), predecessor to the present-day NASA (National Aeronautics and Space Administration), designed entire families of airfoils that were based on different combinations of camber and thickness distributions.

AIRFOILS

One of NACA's first wind tunnels, built around 1920, was called a pressure tunnel. The air was pressurized to 300 psi, which ran the density up by a factor of 20. Therefore, with a $1/20$-scale model, they could get conditions comparable to what the full-scale airfoil would achieve at sea level pressure (14.7 psi). This enabled them to do a better job of designing airfoils. The drawback of the pressure tunnel was that it had a lot of turbulence and would not have been suitable for testing laminar-flow airfoils.

NACA recognized this disadvantage and built a full-scale tunnel in the early 1930s; later, they built a low-turbulence pressure tunnel, where a lot of the laminar-flow work was done.

NASA Langley in Virginia has a full-scale wind tunnel that is 30 × 60 ft. (Fig. 1.1.) NASA Ames in California has one that is 40 × 80 ft and is being made even larger.

1.1. The canard design of the VariEze is being studied in NASA's 30-by-60-foot wind tunnel. *(NASA)*

One of NACA's early creations was the "four-digit" series, including the 2412 airfoil. This airfoil was used on Cessnas, ranging from the Airmaster of the 1930s to the thousands of 152s, Skyhawks, and Skylanes still being cranked out (in addition to the earlier 140s, 150s, 180s, and some of the 200 Series). Other aircraft utilizing the four-digit series include the Navion, Meyers 200, Lake, Beagle 206, Jodel Robin, Wilga, and Zlin. There are far more planes flying with this 50-year-old design than with any other airfoil. (Fig. 1.2.)

Figure 1.2 shows cross-sections of airfoils that are described below by Dan Somers of NASA:

NACA 4415 is one of the four-digit series—the first series designed by NACA in the early 1930s. This turbulent-flow airfoil has relatively good performance at Reynolds numbers below 1 million.

NACA 23015 is an example of the five-digit series, turbulent-flow airfoils, which were designed to produce higher maximum lift coefficients by placing the camber far forward. The higher lift was achieved at the expense of sharp stall characteristics.

NACA 63_2-215 is a six-digit series, laminar-flow design. These airfoils produced lower drag than the turbulent-flow airfoils (assuming a clean wing), but their maximum lift capability was not very good.

LS(1)-0417 is the new designation for the GA(W)-1. A turbulent-flow airfoil, it produces higher maximum lift coefficients than the five-digit series. However, it is not a particularly good airfoil at Reynolds numbers below 2 million. It also has high pitching moments, resulting in high trim drag, and high hinge-moment coefficients, which can lead to aileron float. Furthermore, the airfoil tends to have high leakage through flap and aileron gaps, causing drag.

LS(1)-0417 Mod relieved some of the problems of the early LS (Low Speed) series. The camber was moved forward, increasing maximum lift. Also, it has lower pitching moments, but the stall is sharper. This airfoil tends to be quite good at Reynolds numbers as low as 1 million.

NLF(1)-0416 is the Natural Laminar Flow Airfoil, designed to combine the high lift of the LS series with the low drag of the six-digit series. This goal was achieved; however, because of the aft camber, the airfoil has very high pitching moments.

NLF(1)-0215F is a new airfoil incorporating a simple flap that can be deflected up 10° in cruise, as illustrated. This configuration reduces drag to half that of the LS series.

Turbulent Flow vs Laminar Flow. The four-digit series is a so-called turbulent-flow design. Airfoils are considered to be either turbulent flow or laminar flow, which is a little misleading, because both types produce turbulent flow—and some turbulent-flow airfoils may also produce laminar flow! Stay with me—I'll explain.

Turbulent-flow airfoils may have laminar flow on the top or bot-

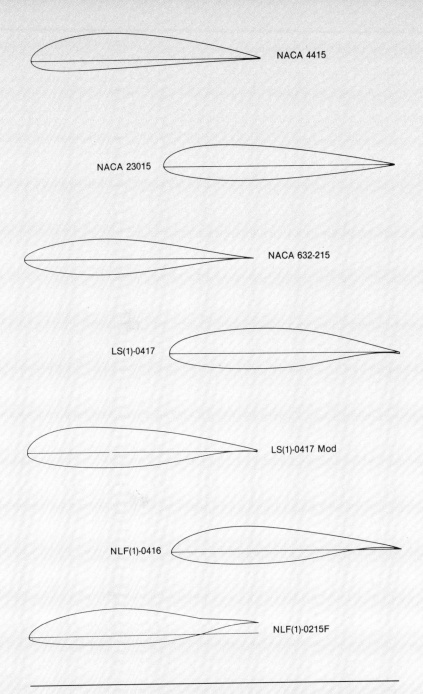

1.2. Cross-sections of some typical airfoils. (*NASA*)

tom surface. For an airfoil to be considered laminar flow, it must achieve that type of flow over at least the first 30 percent of each surface. Laminar flow is defined as a flow in which the air particles move in parallel layers, each having a constant velocity and a motion relative to its neighboring layers.

The region close to the airfoil surface is called the boundary layer; there, the effects of viscosity are predominant. This layer encounters shear stress due to skin friction, and when it separates from the airfoil, the airfoil stalls. So the viscous characteristics of the airfoil close to the surface are really what give the drag and stall behavior to the airfoil. The designer seeks to shape the airfoil in such a way as to control these viscous effects. At the leading edge of a typical general aviation airfoil, the boundary layer is only $1/32 - 1/16$ in. thick. At the trailing edge, it could be as much as 2 in. thick, depending on the airfoil and the angle of attack. When it separates, the boundary layer gets quite large, and a lot of buffeting takes place.

At the surface of the wing, the molecules of air are trapped on the surface and are not moving at all. Layers above the surface are moving at a progressively faster rate, until, at a height of about $1/16$ in. above the wing, the velocity of the molecules has exceeded the flight velocity. To use a specific example, if the wing is moving through the air at 100 fps, the boundary layer may be moving at 200 fps (since there is lift-causing accelerated flow), which slows down to zero at the surface. That causes shear, or friction drag. As the air progresses toward the trailing edge, the flow becomes turbulent, due to disturbances on the wing or in the air itself. This causes an increase in friction drag; therefore, turbulent-boundary layers have more drag than laminar-boundary layers.

With a laminar-flow airfoil, the designer tries to keep the laminar flow intact as far back on the wing as possible. However, not all wings boasting a laminar-flow airfoil actually produce as advertised. Take a nice, clean laminar-flow design, clutter it up with too many protruding rivets, overlaps, or manufacturing imperfections, and there goes the laminar flow. Also, if the skin is thin, it might deform in flight, altering the flow.

Even on a beautifully made wing, laminar flow can be destroyed by an accumulation of bugs. So if you fly with a laminar-flow wing, keep it clean.

The four-digit series airfoils mentioned earlier are described as having reasonably good lift characteristics and drag that is fairly low for turbulent-flow airfoils. They perform well at relatively low Reynolds numbers.

What's a Reynolds Number?

A Reynolds number is equivalent

to air density multiplied by airspeed, multiplied by the wing chord, divided by viscosity. As a rule of thumb, if you're flying 100 mph at sea level, the Reynolds number is equal to 1 million for every foot of wing chord. Therefore, if you're skimming along the deck at 200 mph, and your plane has a wing chord of 5 ft, your Reynolds number is 10 million. As you go up in altitude, the Reynolds number decreases along with the density of the air.

If two airfoils of different sizes are operating at the same Reynolds number, the flow about them will be geometrically and aerodynamically similar, and the maximum lift and drag coefficients can be compared directly. This helps designers to extrapolate small-scale wind-tunnel test results to full-scale applications.

Designers look at the Reynolds numbers at which a plane will be operating when evaluating the lift and drag characteristics of the airfoils under consideration.

The Search for Improved Max Lift. After the four-digit series, NACA came up with a five-digit series, in which the camber was moved considerably forward. This provided an improved maximum lift coefficient, along with lower pitching moment and, therefore, lower trim drag. However, it is a turbulent-flow airfoil, and the drag is somewhat comparable to the four-digit series.

The best known of the five-digit airfoils is the 230 series, which is used by Beech on their Bonanza, Baron, Duke, and King Air models, as well as by Cessna on the Caravan, the 300 and 400 models, and the Citation I and II, and by Piper on the Malibu.

The 230 series airfoil has a sharper stall characteristic than other airfoils. This does not mean a wing made up of this series will necessarily have a sharp stall; the stalling characteristics of a wing can be tailored by its planform shape, taper, twist, thickness distribution, stall strips, strakes, and fences. The designer's goal is generally to come up with a wing whose stall will occur first at the root and then move progressively outward, so aileron control can be maintained.

A point of interest: airfoils with leading edges that "hang on" and provide good stall behavior may have less desirable spin recovery characteristics. For this reason, the cambered leading edge modification Cessna put on the 172, 180, 182, and 206 was not given to the 150/152.

Enter Laminar Flow. Even as far back as the late 1930s, NACA had concluded the only way to lower drag—and thus attain higher cruise speeds—was through laminar flow. So they designed and tested a number of laminar-flow airfoils, of which the most widely used is the "six" series, whose designations began with the numeral 6. (Example: 63_2-215.) These designs are found on the Piper Cherokee line, Mooney, Beech Sundowner, Sierra, and Duchess, Learjet, MU-2, and others.

Achieving laminar flow can reduce airfoil section drag by as much as 40 to 50 percent, in comparison with four- or five-digit airfoils of the same thickness. However, as mentioned before, the wing-surface contours must be sufficiently smooth to attain laminar flow. Since the six series does not have very high max lift coefficients, it's not a good choice if your wing does not achieve laminar flow, because you're giving up lift without getting the benefit of reduced drag.

Also, the six-series airfoils have leading-edge separation. Therefore, if the leading edge of one wing didn't quite match the leading edge of the other wing, you could get unequal stall characteristics and roll off on one wing during a full stall. A typical fix by an aircraft manufacturer was to reposition the leading-edge stall strips to obtain a symmetrical stall. The newer analytic methods, using the computer, have allowed designers to change the upper surface of the leading edge and make it much more docile, so instead of having the stall break at the leading edge, the stall would go from the trailing edge forward. This produces a gentle stall rather than a sharp break.

To sum up, at this point NACA had designed turbulent-flow airfoils that produced good lift but fairly high drag, plus laminar-flow airfoils that had the potential (on a clean wing) for low drag but not high lift.

Then, in the early 1950s, NACA got out of the low-speed airfoil business to concentrate on transonic and supersonic designs. By the mid-1960s, the agency had changed its name to NASA and was working on the supercritical wing, in the Mach 0.8 regime, for transport category airplanes. The camber of this airfoil is located primarily near the trailing edge. It is not suitable for low-speed aircraft, but a derivative was developed for that purpose — the GA(W)-1 airfoil.

In designing the GA(W)-1, no attempt was made to achieve laminar flow because of the tendency of most aircraft manufacturers to use round-head rivets, thin skins, and widely spaced ribs, which are not conducive to laminar flow. Instead, the goal was a turbulent-flow airfoil with improved max lift coefficient, and this result was achieved.

However, because the airfoil had an aft camber, somewhat like the supercritical wing, it produced high pitching moments and high aileron hinge moment coefficients — the latter caused cable-actuated ailerons to float upwards at cruise. So NASA designed a modified version of the GA(W)-1, with the camber moved forward. This improved the lift and decreased the pitching moment, at a cost of somewhat less docile stall characteristics.

The GA(W)-1 airfoil is used on the two "new generation" trainers, the Piper Tomahawk and Beech Skipper. Interestingly, the stall/spin characteristics of these two planes are quite different. The Tomahawk has a sharply-breaking stall and can be made to spin quite easily, while the Skipper is more docile and reluctant to spin. This is a further indication the performance of a given airfoil depends largely on the design and construction of the *entire* aircraft.

NASA followed the GA(W)-1 with a GA(W)-2, and at the risk of inundating you with numbers, I must tell you that the GA(W)-1 is now officially called the LS(1)-0417 and the GA(W)-2 is the LS(1)-0413. (Incidentally, the last two numerals on all NACA/NASA airfoils refer to their percentage of maximum thickness relative to the chord.)

At Last—High Lift and Low Drag. In the mid-1970s, NASA set about designing a series that would combine the high lift of the LS series with the low drag of the six series, and this was accomplished with two new laminar airfoils, NLF(1)-0416 and NLF(1)-0215F.

One problem that had to be overcome was the bug situation; if a plane with the six-series airfoil picked up bugs, its landing speed could actually be increased. The designer of the new airfoils, Dan Somers, has this comment:

> The nice thing about them is that even if you don't get laminar flow, because the wing is poorly made, or you get bugs or rain, you'll still get high lift—and the drag will be no higher than if you had a turbulent flow airfoil of the same thickness. But if you *do* get laminar flow, then you'll have a substantial drag reduction.

The NLF(1)-0215F utilizes a flap that can be deflected *upwards* 10° at cruise, much like high-performance sailplanes. The airfoil is designed for the new generation of high-performance single-engine planes that are expected to cruise at 300 mph at 25,000 ft. The new pressurized Mooney 301 will have a similar airfoil. However, it will not have the upward-deflecting flap. Mooney designed their own natural laminar-flow airfoil and utilized a very substantial Fowler flap—90 percent of span—for low-speed takeoff and landing characteristics.

What Lies Ahead? Airfoil design has been made much less laborious by—you guessed it—the computer. In the 1960s, Professor Richard Eppler of the University of Stuttgart began creating computer programs for airfoil design analysis. These programs have been further developed by Eppler and Dan Somers, working together at NASA Langley. Eppler has been responsible for a number of airfoil designs, including Burt Rutan's Defiant, and participated in the design of the Mooney 301.

In precomputer days, a person using a mechanical calculator would take one week to come up with a single pressure distribution. The same work can now be done on a computer in one second. This frees the designer from having to select from existing airfoils and accepting compromises in the process.

With the computer programs now available, airfoils can be designed for specific airplanes, taking into account the aircraft's projected weight, lift coefficient at various speeds, and flap design.

It is to be hoped the aircraft manufacturers will make increasing use of these opportunities.

2. Flaps, Ailerons, and Spoilers

What to look for in a flap...Plain and Frise-type ailerons... Spoilers that do away with ailerons.

EVEN the most efficient wing has its airfoil adjusted many times in flight. This is done with movable control surfaces—flaps, ailerons, and spoilers—that improve takeoff and landing performance and enable us to make turns.

Just as there are different designs for airfoils, so are there varying designs for these movable control surfaces.

Flaps. Four basic types of flaps are used on general aviation aircraft: the plain flap, the simple slotted flap, the rearward-action flap, and the split flap. (Fig. 2.1.)

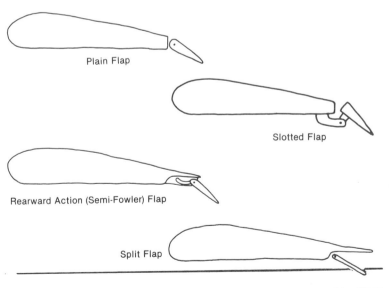

2.1. Flaps

THE PLAIN FLAP. This is just what its name implies—a simple structure that is hinged like a door to the trailing edge of the wing and has an uncomplicated down-and-up movement. The plain flap is found predominantly on older aircraft, as well as on Beech's current trainer, the Skipper.

THE SIMPLE SLOTTED FLAP. Like the plain flap, this flap has a fixed pivot point. When it is deployed, a slot, or gap, opens up. This is designed to improve the airflow over the flap's upper surface. All of the current Piper singles plus the Seminole and Seneca have this flap, as do the Beech Sundowner and Sierra.

THE REARWARD-ACTION FLAP. This is sometimes referred to as a semi-Fowler flap, or a flap with Fowler movement. To best explain this, let me start with a description of the Fowler flap. Whereas other flaps constitute part of the wing surface even when retracted, the Fowler is nestled under the wing in its retracted position. As it is deployed, it moves rearward as well as downward, adding to the effective wing area and providing a greater lift coefficient. Heretofore, the full Fowler flap has been considered to be too complex for use in singles and light twins, having been reserved for some of the larger business twins and commercial aircraft. However, it is being used on the Mooney 301 pressurized single.

Many general aviation flap systems employ tracks that provide some aftward movement, although less than that of the Fowler system. The main purpose of this rearward movement is to minimize flow separation of air that passes through the slot and over the flap. Cessna uses this type of flap on all their singles and on the Crusader. It is also found on Beech aircraft, from the Bonanza on up to the King Air, and on the Piper Malibu.

THE SPLIT FLAP. Like the Fowler flap, the split flap is nestled under the wing, but has a fixed hinge point, no rearward movement, and no slot effect. It is generally thought of as providing little lift and a lot of drag. However, Cessna's Manager of Advance Design and Systems Research, Dave Ellis, has this to say on the subject: "I think you'll find it's a pretty fair lifting flap—better than most people realize. It isn't too much worse than a slotted flap for lift, but it does give considerably more drag."

The split flap is used on the Cessna 310, 335, and 340 aircraft. Some of the 400 series airplanes also have the split flap, while others utilize the Fowler system.

Generally speaking, a flap will provide more lift than drag at the lower flap settings, and more drag than lift at the higher settings. For some aircraft, the manufacturers call for zero flap deflection on takeoff; for others, partial-flap settings are specified, depending on the

FLAPS, AILERONS, AND SPOILERS

type of takeoff. Many handbooks stand mute and let the pilot figure it out for himself.

Another consideration is the flap-extension speed, which, like gear-extension speed, directly relates to your ability to slow the plane for descent and pattern work. Many aircraft have an approach setting of 15° or so that can be used at a fairly high speed; this helps get the plane slowed down to the airspeed indicator's white arc, at which point full flaps can be progressively deployed. An approach setting is also useful in obtaining a combination of nose-low visibility and reduced speed in the traffic pattern.

Then we come to the effectiveness of the flaps as drag-producers, to provide steep approaches without airspeed buildup. Here again, there's considerable variation from model to model. The flaps on some airplanes, such as the Grumman-American singles, appear to have been put there to give the pilot something else to play with; in short, they are not very effective. Conversely, the legendary barndoors on the Cessna singles will give you a nice express-elevator ride. In fact, they became a bit too much of a good thing when deployed to 40°, blanking out the propwash and reducing rudder control considerably during a full-flap go-around. Cessna's balked landing procedure called for partial flap retraction after application of full power, but some pilots got behind the airplane. As a result, current Cessnas have a max flap setting of 30°.

Yet another consideration is the flap extension system. Some are electric, some are hydraulic, and some are just plain mechanical. The mechanical "crowbar" systems, such as Piper uses on their Vero Beach models, have the advantage of being practically maintenance- and failure-free—one less thing to worry about. Also, they enable you to control the retraction speed. Sometimes you want to get your flaps up right now, such as on touchdown in gusty conditions. Other times you may want to milk them up *verrry* slowly, during a short field takeoff when the trees are coming at you. The main disadvantage of the mechanical system is that it requires more muscle, especially on a fairly big airplane like the Seneca.

Of the powered systems, the most convenient ones are those whose controls have detents at the various settings. It's rather distracting to be holding a spring-loaded flap control and watching a gauge on final when you may have other priorities.

Another problem associated with having a flap handle on the panel is that you could mistake it for the gear handle. If you did this on landing, you would then have, as airline captain and writer Barry Schiff once put it, a rollout that was much shorter than anticipated. The potential for this error is especially acute in all the Bonanzas and Barons made prior to 1984. For years, Beech persisted in placing the landing gear switch to the *right* of the flap handle when it has been to the *left* on every other modern retractable, including their own Sierra and Duchess. Finally, in 1984, Beech came out with a redesigned panel for the long-bodied Bonanzas and the Barons, with the flap switch on the left side. Presumably, they'll get around to making the same fix on the short-bodied Bonanzas before long.

Ailerons. The ailerons in use today are the plain aileron and some sort of Frise-type aileron—usually modified from the original Frise design. (Fig. 2.2.)

The plain aileron, like the plain flap, is a simple affair that is often connected to the wing by piano hinges.

The pure Frise aileron has a setback hinge and a sharp nose. It is so arranged that when the ailerons are deflected, the up aileron's nose projects down in to the airstream, but the down aileron's nose does not project up into the airstream.

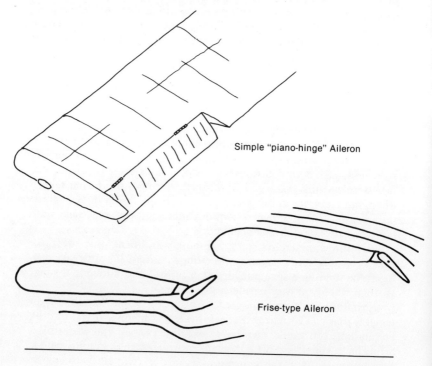

2.2. Ailerons. Up aileron's nose projects down into the airstream. Down aileron's nose does not project up into the airstream.

FLAPS, AILERONS, AND SPOILERS

There are two purposes to this design. The first purpose is to help counteract adverse yaw. When you put your plane into a bank, it tends to swing in the direction of the ascending wing. This happens partly because that wing's down aileron is producing more drag than the descending wing's up aileron, and partly because the lift vector of the rising wing has a rearward tilt, while the lift vector of the lowering wing has a forward tilt. When the nose of the up aileron projects into the airstream, it creates drag, slowing the faster wing somewhat. (Trevor Linton-Smith of Piper disputes this theory, stating that the aileron drag has little to do with adverse yaw, which can best be handled by giving the plane ample directional stability via the vertical tail.)

The second function is to aid in the deflection of the ailerons. As the nose of the up aileron encounters the airstream, it is pushed by the force of that airstream, thus reducing the control forces the pilot must exert.

The problem with this system is if a pair of Frise ailerons is not perfectly matched, aileron snatching can occur, wherein the control surfaces deflect faster or further than the pilot wishes.

The solution lies in a modified Frise aileron with a rounded leading edge and a hinge arrangement that projects the nose into the airstream in both the up and down positions. This doesn't do much for adverse yaw, but it does decrease control forces without aileron snatching.

The Frise-type ailerons are, by definition, aerodynamically balanced. Ailerons must also be balanced statically, with internal or external weights, in order to suppress flutter tendencies.

Another way of decreasing control forces is to increase the amount of control wheel throw from lock to lock. This was done with the Mooney 201 and 231, because their faster speed made aileron forces higher than those of the older Mooneys.

Yet another way of decreasing adverse yaw is to provide a deflection differential—that is, design the aileron to go up at a greater angle than it goes down, thus reducing the drag of the down aileron.

Plain hinged ailerons were part of the simple design philosophy behind the original rectangular-wing Piper Cherokees, which, despite their many virtues, were not renowned for crisp roll control. In fact, these ailerons were quite inadequate for the Seneca I, so the Seneca II was given modified Frise ailerons, as were the models with the semi-tapered wing (i.e., the later Arrow, the Archer, Dakota, Saratoga, and Seminole).

The Tomahawk has plain ailerons. So does the Mooney and the Beeches up through the Baron, except for the Duchess, which has a Frise type. Some of the ailerons are slotted, and those on the Bonanza and Baron have some internal aero-balance.

Cessna puts plain ailerons on the 152, 172, 182, and 185. The 206, 207, and 210 aircraft have modified Frise ailerons, while the 208 and 300 and 400 series planes are equipped with round-nose ailerons that have back-set hinges for aerodynamic balance.

Spoilers. Spoilers can be used either for descent control or for roll control—augmenting or replacing flaps for descent and ailerons for roll. Except for sailplanes, where spoilers are the primary descent control, they have had little usage by general aviation aircraft manufacturers. Some mod shops are installing them for the aftermarket. For example, Precise Flight offers speed brake systems for a number of singles. And R/STOL, Inc. (formerly Robertson Aircraft Company) is installing spoilers and full-span flaps on the Seneca I and Bonanza V35 and A36 models. (Fig. 2.3.) For more details on these systems, see Chapter 14.

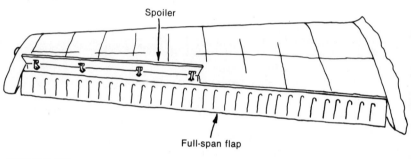

2.3. Spoiler

When spoilers are used for roll control, the area that previously had been taken up by ailerons can be devoted instead to larger-span flaps. For example, the Mitsubishi MU-2 has spoilers and 90 percent-span flaps. The Mooney 301 will have spoilers *and* ailerons. Here's why, in the words of aircraft designer Roy Lopresti:

The spoilers are there primarily because we wanted full-span flaps for short takeoff and landing distances. We have small "feeler" ailerons in addition to the spoilers, just to be sure that the roll control works exactly the way we want it to work.

Ailerons are not as effective as properly designed spoilers, particularly at low speeds. On the other hand, if spoilers are not properly designed aerodynamically, they can have certain peculiarities, such as dead bands or force gradient variations at deflections of about 2 to 6 degrees.

To summarize, spoilers can give better roll control than ailerons, but it takes a higher level of technical competence to attain the maximum benefit.

FLAPS, AILERONS, AND SPOILERS

The Cessna 208 Caravan has a mix of ailerons and slot-lip spoilers.

On the other hand, a Robertson spokesman claimed that their spoiler system works perfectly well without the help of ailerons, adding that spoilers give more positive roll control at slow speeds, where the reduced airflow makes ailerons less effective.

Assuming a conventional flap and aileron arrangement, how does the designer decide on the relative sizes of each control surface? In fact, how is the size and shape of the entire wing determined? Here is the sequence of events at Beech, according to one of their design engineers:

When they size a wing for a new airplane, they start with the plane's required landing distance. Then they determine how much power it will take to get the plane off in the prescribed landing distance with that size wing. Then they'll look at the resulting climb and cruise figures and tweak the various considerations to arrive at a good balance.

Because of the importance of roll control power, the aileron span helps to establish the size of the flaps. But sometimes the chord of the aileron is increased so its span can be reduced, to allow for additional flap span. In other words, the real estate on the wing can be parceled out in a number of ways, depending on the airplane's desired performance characteristics.

What about the planform of the wing? According to Dan Somers of NASA, to get the minimum induced drag, the designer would like to have an elliptical lift distribution, which he could get with a wing that had curved leading and trailing edges, and no twist in it. Some older airplanes, like the Spitfire, had this type of wing. The problem was that they were hard to build.

The latest sailplanes have triple-tapered wings, in an attempt to approximate an ellipse. The closest thing you'll find to this in a powered light plane is the semi-tapered wing. Piper went to this in a big way when they gradually abandoned the rectangular, or "Hershey bar" wing, in the Cherokee series. Now all of the current Cherokee derivatives, except the Seneca, sport semi-tapered wings, and most people feel their roll control has been notably improved.

3. Tails

The tail alphabet: T, V, and Y... Are the fancy tails just sex symbols?... Tail first with the canard.

AIRPLANE TAILS come in a variety of shapes and sizes. The designer attempts to provide tail surfaces large enough to give effective control without undue weight and drag penalties. Sometimes he winds up on the stingy side, and you find yourself running out of rudder in a crosswind landing, or elevator in the flare. Location of the horizontal surfaces with respect to propwash and downwash is another consideration, hence the appearance—and sometimes disappearance—of the T-tail.

The majority of the piston-engine planes have the so-called conventional tail, with the horizontal surfaces extending from the tailcone of the fuselage. Let's take a look at some of the less conventional designs.

The T-tail. T-tails are found on the Beech Skipper and Duchess, as well as the larger King Airs. Cessna has a T-tail on the Citation III, but not on the Citation I and II. Piper put T-tails on late models of the Lance and Arrow, the Seminole, and the Cheyenne III. (Fig. 3.1.) In 1977, a company called Anderson Greenwood brought forth a T-tail single called the Aries T-250. The plane never got into full production, because Anderson Greenwood had the misfortune to get financially involved with the terminally ill Bellanca Aircraft Corp.

Piper's first flirtation with the T-tail turned into an embarrassment. Two years after the Lance had been in production, Piper decided to improve the six-place retractable by replacing its conventional derriere with a T-tail and calling it a Lance II. Improvement? The published takeoff over the 50-ft barrier went from 1620 ft for the Lance I to *2240* ft for the Lance II. Furthermore, many pilots experienced over-controlling when breaking ground on takeoff and difficulty in getting a consistently good landing flare.

With proper piloting technique, these problems could be handled, but people stopped buying Lances in droves. Piper experimented with a larger horizontal stabilator, but this would have required further beefing up of the vertical stabilizer, and the added

3.1. The 1983 Piper Turbo Arrow. *(Piper Aircraft Corp.)*

weight would have been too big a burden. So Piper bit the bullet and changed the Lance into a Saratoga SP. It featured two major improvements: a semi-tapered wing and a low tail. (The fixed gear version, formerly called the Cherokee Six, later the Six 300 and currently the Saratoga, had the good fortune to escape the transitory T-tail.)

A year after introducing the Lance II, Piper brought out the Arrow IV, which boasted, of all things, a T-tail. (There must have been a lot of crossed fingers in Vero Beach and Lock Haven.) The change worked out reasonably well on the shorter-fuselaged Arrow, in that there was no degradation in performance. There was no real improvement, either, except for those who like the look of a T-tail.

There is a price to be paid for that look. Disadvantages of the T-tail include a weight penalty, a more complicated control mechanism, and the difficulty, or downright impossibility, of preflighting the horizontal stabilizer without a ladder.

Even the manufacturers of the big iron sometimes have difficulty deciding whether to go the T-tail route. There's a story that when Boeing was designing the 757 and 767 aircraft, they had one design team working with a T-tail and one team developing a low tail. The first two customers, asked for their preference, opted for the low tail, whereupon the T-tail team was promptly disbanded.

A Halfway Measure. Rockwell's 112, 114, and 700 aircraft, along with Cessna's 303T Crusader and Citation I and II, sort of split the difference between a conventional tail and a T-tail. Their planes have a cruciform tail, with the horizontal surfaces located partway up the vertical stabilizer.

The V-tail. This is probably the most distinctive and long-lasting trademark in the history of general aviation. I refer, of course, to the tail of the Model 35 Bonanza, which has been in production since 1947. (Fig. 3.2.)

For many years, the Bonanza used only the V-tail. Then, in 1960, a lower-priced companion airplane was introduced, to compete with the Piper Comanche. The new Beech plane was given the Model 33 designation and was called a Debonair. It was very similar to the Bonanza, except it had a less powerful engine, options that were standard on the Bonanza, and a straight tail. Like many poor relations, the Debonair was not well accepted—either by the public or by Beech's retailing organization—so the plane was gradually upgraded over the years until, in 1968, it was let out of the kitchen and allowed to use the Bonanza name. Since then, there have been two Bonanza models that are identical in all respects except for the tail. And 1968 was truly a Bonanza year, in that an aerobatic version of the 33 was introduced, along with a stretched Bonanza, the Model 36, that also was given a straight tail.

The V-tail, sometimes informally known as the butterfly tail, has two movable control surfaces called ruddervators. By means of a rather intricate coordinating mechanism, the ruddervators deflect either in the same direction or in opposite directions to perform the dual functions of rudder and elevator.

Beech originally went to the V-tail to achieve the following benefits: less weight, less drag, lower manufacturing costs, and improved spin recovery characteristics. However, anyone comparing the factory specifications of the V35B and the F33A might be understandably confused. The V35B is shown to have a standard empty weight 19 lb less than that of the F33A, but the published performance figures and list prices of the two planes are identical.

3.2. A Robertson-modified Bonanza, with its distinctive tail. (*Jim Larsen*)

Both of these airplanes tend to "hunt," or wag their tails, in turbulent air. Some people say that the V-tail models hunt to a greater degree than the straight-tail Bonanzas. Others claim all short body Bonanzas perform equally in this respect. (Everyone seems to agree the long body Bonanzas are more stable than their shorter kin.)

Another claim against the V-tail is far more serious. It was given widespread attention in the Feb. 1, 1980, issue of *The Aviation Consumer*, whose lead article bore the graphic title, "The V-Tail Bonanza—Breaking of a Legend." I won't try to capsulize the 12-page article in this chapter, but suffice it to say the author, Brent Silver, enumerated statistics indicating the V-tail Bonanzas had a much higher failure rate than the straight-tail models, and he suggested a number of possible tail-related design deficiencies. Silver's accusations reached a much wider audience when the popular *60 Minutes* televised its version to the world at large, using such casual phraseology as "The record shows (that the Bonanza) falls apart now and then."

On June 12, 1980, Beech issued a safety communique in rebuttal, stating, among other things, "The Bonanza does not 'fall apart,' but it can be pulled apart in the air if, by a combination of excessive speed and pilot action, loads are imposed beyond the design limits of the structure."

Beech went on to show a high percentage of these accidents took place in IFR (instrument flight rules) conditions, and many of these involved pilots who were not instrument rated. But why the high ratio of V-tail to straight-tail Bonanza failures? Again, Beech's answer is too lengthy to quote in detail, but the following excerpt is interesting:

> In the opinion of Beech engineers, the V-tail aircraft retains full maneuverability deeper into the spectrum of flight at excessive speed than does the conventional tail. This is not to say that the conventional tail becomes completely rigid or unmaneuverable. Rather, the conventional tail tends not to provide or retain sufficient maneuverability to avoid ground impact at excessive speeds. The V-tail does, but at the cost of imposing loads on the airframe that are beyond its design strength.

Translated into plain English, this appears to say: If you get into an uncontrolled excessive-speed situation, you're more likely to bore a hole in the ground with the conventional-tail Bonanza than with the V-tail model, whereas with the latter you'll have a greater chance of pulling up, but the airplane may come apart in the process.

Personally, I'd be hard put to choose between these alternatives. Instead, I'd recommend a third choice, namely, don't get into an excessive-speed situation in the first place. How? First, stay out of flight conditions you can't handle, even if it means missing a business appointment or not getting home in time for dinner. But if you do get in over your head, follow Beech's advice and get the landing gear down. This will help prevent extreme speed buildup while you get yourself and the airplane under control again. Even if you've exceeded gear extension speed, pull that little round handle and sacrifice the gear doors, if necessary, to save your life.

Also, if you own a V-tail Bonanza, you might consider one of the modifications that are designed to strengthen the tail section. See Chapter 14 for further details.

The future of the V-tail Bonanza is in question. At this writing, Beech has suspended production of that legendary plane—because there are no orders for it—and is manufacturing only the straight-tail models.

The Y-Tail. This one looks like the Bonanza V-tail with a vertical fin and rudder stuck on the bottom for good measure. It's to be found on the Lear Fan. Why? Well, according to its would-be manufacturer, "Wind tunnel tests at the California Institute of Technology compared a variety of conventional and experimental tail designs. The Y proved best for stability and control."

It has also been suggested by more cynical sources that the Y-tail design will keep the big pusher prop from striking the ground. But although the tip of the down-pointing fin itself appears to be perilously close to the tarmac when the Lear Fan is at rest, the manufacturer maintains it won't hit bottom unless the centerline of the aircraft rotates to 11°, which is double the landing flare attitude.

The Mooney Tail. The tail found on the Mooney M 20 series is unique. First of all, compared to the swept-back tails that are virtually the standard of the industry, the Mooney tail looks as if it had been installed backwards by a careless worker. The leading edge of the vertical surface is straight, while the trailing edge is slanted. But that's not all. When the pilot cranks in a change in pitch trim, the entire tail rotates. For example, if you move the trim wheel back, the vertical stabilizer and rudder will lean forward as the horizontal stabilizer and elevator tilt. The purpose of this is to keep the rudder hinge line vertical in the slow flight trim condition, and thus maximize the effectiveness of the rather slender rudder.

There is a potential side effect to this arrangement. Some years ago, I owned a Mooney Super 21. Its ADF (automatic direction finder) "clothesline" antenna ran from the cabin roof to the tail, and, of course, was slack when the tail was in the forward, or nose-high, trim position. More than once I picked up my plane from a maintenance shop, only to find that an assiduous mechanic had neatened things up by replacing the "loose" antenna with one that was tight. The thought of belatedly discovering I was wired into a nose-high attitude gave me further respect for both the preflight checklist and the premaintenance A&P (airframe and powerplant mechanic) briefing.

The Canard. Burt Rutan, the wizard of Mojave, California, has been responsible for a string of homebuilt aircraft designs that have provided astonishing performance and a high degree of stability and safety—all utilizing the canard concept. Beech has taken hold of the idea with its Starship I, a corporate turboprop that looks something like a giant VariEze.

A canard is a horizontal stabilizer and elevator mounted ahead of the main wing. One getting his first glimpse of an airborne VariEze, with its small forward wing, large rear wing, and pusher prop, might be forgiven if he thought the plane were somehow flying backwards.

Rutan's aircraft do not stall in the usual sense, nor do they spin. Here's why, as the designer explains it:

> What we do is naturally limit the angle of attack which the pilot can command with the elevator by designing the two wings so that the canard progressively stalls in a planned progression. Once you reach the angle of attack at which it's no longer useful to go to higher lift, and going to higher lift would stall the wing, then the canard (forward wing) progressively stalls so that it can't force the (main) wing to a stall angle of attack.
>
> That gives the airplane what amounts to a strong, stable break in the pitching moment curve. What that means is, once you reach a high but still-safe angle of attack, it's impossible to force it higher.

Rutan's canard designs also trim away a lot of the structure necessary in conventional aircraft. We'll go into this and other canard benefits when we look at the Defiant in Chapter 14, "Airplanes of a Different Sort."

One disadvantage to the canard design: it reduces visibility, which is particularly noticeable when the pilot is trying to get some perspective in the landing flare.

The Stabilator. The majority of the planes we fly have their horizontal tailfeathers in the form of a fixed stabilizer and a moving elevator. However, some aircraft, notably the Piper Cherokees and the Beech Sundowner and Sierra, are equipped with an all-moving horizontal control surface called a stabilator. To find out if there are any significant advantages to this design, I talked to Trevor Linton-Smith, chief of Aerodynamics at Piper's Vero Beach factory.

He told me the designer will elect to use a stabilator if he is designing an airplane where the CG (center of gravity) envelope is fairly far forward. The stabilator will usually provide more tail power than a conventional elevator for the same amount of area. Actually, the stabilator requires about 25 percent less surface area, but the surface has to be mass-balanced to provide good handling characteristics and to meet flutter requirements. Therefore, there is an area advantage, but no weight advantage.

Because the stabilator has to be mass-balanced 100 percent, the stick forces cannot be tailored as they can in a conventional elevator.

These factors have some effect on handling quality. If you are transitioning from a plane with a conventional elevator to one with a stabilator, you may tend to over-control in the flare, because you'll get more reaction from a given amount of stick motion than you are accustomed to.

Rudder Trim Systems. The simplest kind of rudder trim is a fixed metal tab, which is ground-adjustable with the aid of a pair of pliers. Obviously, this is set just for cruise condition, and doesn't help overcome torque and P-factor during climb. A little surprisingly, some of the sophisticated (and expensive!) singles such as the Mooney and Bonanza do not have flight-adjustable rudder trim. This can be particularly annoying on a turbocharged model, when you've got to hold right rudder for a climb to 25,000 ft.

There are two ways to provide trim that can be adjusted in flight. The smoothest and most precise way is through the use of a movable trim tab. This is found on all twins, because it is the only method of satisfying single-engine performance requirements. It is also used on some singles.

The alternative is a spring contrivance that simply exerts tension on the rudder pedals. The Maule has the plainest of these devices. Prior to takeoff, you pull a T-handle, which provides spring action on the right pedal. When you level out, you retract the T-handle. Versions found in other aircraft permit variable amounts of left and right trim, but sometimes you have to play with the rudder pedals to get it just right. The designer will opt for the spring instead of the nicer movable trim in order to save parts and sometimes to avoid the possibility of tab-induced flutter.

Rudder/Aileron Interconnect. A number of planes have this arrangement, which is an aileron-centering spring that is activated when the rudder is deflected. Its purpose is not to help pilots make coordinated turns, but rather to pass the lateral static stability test for certification.

This calls for slipping the airplane on a heading, which, of course, requires the use of rudder and aileron. Then the aileron is released, after which the low wing must show a tendency to come up. If the aerodynamic hinge moments are low and the control system friction is fairly high, the ailerons will not tend to center, and then the low wing won't come up.

Well, we've looked at the shapes and surfaces that make an airplane fly. Now let's get to the power.

4. Engines and Fuel

The practical meaning of TBO...Operating tips that can save you money and save your engine...Water-cooled engines and Wankels...What about autogas?

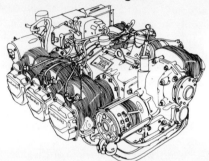

IF YOU'RE FLYING a production plane manufactured within the past decade, chances are it's powered by either a Continental or a Lycoming engine. There used to be a Big 3, the third member of the triumvirate being Franklin, but that company ceased production, and its tooling is now owned by a Polish firm.

Most aircraft engines have enjoyed a good history of reliability. However, a few models have been plagued with rather serious problems, such as cases that cracked, or connecting rods that failed. Other engine malfunctions have been caused by a run of bad magnetos or other accessories made by outside vendors.

Check the Track Record. If you are in the market for an airplane, it's a good idea to look at the track record of the aircraft, its engine, and related accessories. In time, most of the serious problems get cleaned up, often after an AD (airworthiness directive) has been issued by the FAA mandating the corrective action, but you want to be sure the engine that's keeping you in the sky has had the benefit of any necessary updates.

Here are several ways you can get information: *The Aviation Consumer,* published twice a month, is pretty aggressive at reporting airframe and engine problems. Also, you can order an FAA service difficulty report on a particular make and model of aircraft. The SDR is a compilation of all reports of malfunctions and defects submitted to the FAA for up to six previous years by mechanics and inspectors. It can be ordered from the FAA or the Aircraft Owners and Pilots Association's Aircraft and Airmen Records Department, both in Oklahoma

ENGINES AND FUEL

City. The cost of an SDR is upwards of $50, depending on the size of the document. Aero Tech publications will send you a listing and description of the airworthiness directives, service bulletins, and service letters that have been issued on just about any make and model of airplane for under $20.

TBO. In addition, you'll want to know the engine's TBO, or time between overhauls. This figure, usually running somewhere between 1200 and 2000 hr, is set by the engine manufacturer and represents their judgment as to how long the engine should run before it needs to be overhauled. Sometimes a manufacturer will increase an engine's TBO, based on its service experience.

But nowhere is it written that your engine will hum along right up to its TBO and then go to pieces. The actual service life will depend largely on the way it's been operated and maintained. Your powerplant may not make it to its published TBO, or it may still be showing signs of perfect health when that number rolls up on the tach.

In the latter case, you may decide to keep it running until it does get tired. After all, there's that old saying, "If it's still working, don't fix it." If you exercise that option, you should have the engine inspected every 100 hr past the TBO.

However, Continental warns that running an engine past its TBO may be poor economy. One reason for this is that as parts become worn, the *rate* at which they continue to wear *increases*. Therefore, the wear rate will be much higher after, say, 1600 hr than it was at 1000 hr; so if you operate well beyond the TBO, you may be faced with a considerably higher overhaul bill than if you'd had the work done at the recommended TBO point.

There are other "poor economy" practices, such as skimping on oil changes and spark plug replacement, and we'll get into these a little later.

TBO also becomes significant under certain terms of the engine warranty. For example, Continental warranties its engines for 12 mo or 480 hr, whichever comes first. After that period, the crankshaft and crankcase are additionally warranted and the cost of their replacement is prorated to the TBO. So, the higher the TBO, the greater the warranty responsibility of the engine manufacturer.

Now let's look at some other variables.

In some engines, notably those in the 200-hp range, you might find that Lycoming provides four cylinders, while Continental favors six. The tradeoff here is that a four-cylinder engine should cost less to maintain, while the six-cylinder counterpart is likely to run more smoothly.

An Avgas Fuel Crunch? If you're buying an older plane, you'll want to know whether it drinks 80-octane or 100/100-LL fuel. The availability of 80-octane avgas is spotty in many parts of the country, and its manufacture may be discontinued altogether in the near future. In fact, a number of aircraft industry spokesmen predict within the next decade, there will be no aviation gasoline whatsoever. (Jet fuel will still be refined.)

What will we do then? As one avenue, Beech has developed a system that enables cars and trucks to run on liquified methane, and has test-flown a Sundowner on that fuel. A more immediate solution would be the use of auto fuel in planes. Until recently, that was totally illegal, but STCs (supplemental type certificates) are now being made available for a growing number of aircraft models. We'll discuss this in detail later in this chapter.

Getting back to 80-octane engines, some of them have been modified to accept 100-octane fuel more readily. If you have one that has not been so modified, you can carry a can of TCP (Alcor packages it) and add it to your tank when you have to refuel with 100-octane avgas. This will help reduce plug fouling and engine deposits. Your best bet, of course, is to choose a plane whose engine has been designed to burn 100-octane fuel. This includes virtually all of the piston engine aircraft currently being produced.

Carburetor vs Fuel Injection. Some engines get their fuel via a carburetor and others use a fuel injector. There are advantages and disadvantages to each system.

The carburetor is simpler and costs less to manufacture. Also, a carbureted engine is less subject to hot-start problems due to vapor lock. However, it *is* subject to carburetor ice. Since the latter can be a worrisome problem, particularly with certain engines and in certain climates, it is advisable to equip a carbureted plane with a carb temperature gauge and an ice warning light.

The fuel injector is more efficient than a carburetor in that it can meter the fuel more evenly to each cylinder. The legendary hot-start difficulties are not insuperable, and later in this chapter I will pass along a ritual that, according to Continental, works every time. Fuel injection is accepted by the industry as the modern fuel distribution system; the main reason some carbureted engines are still being turned out is cost.

The engine's model number provides a lot of information. Let's take a Continental GTSIO-520-M. The *G* stands for geared drive. The *TS* tells you it's turbo-supercharged. (Lycoming uses just a *T*.) The *I* is for fuel-injected, and the *O* is for horizontally opposed. The engine has 520 cu in. displacement, and the dash letter or letters (sometimes

ENGINES AND FUEL

numbers) refers to a modification or accessory. What the model number does *not* specify directly is the horsepower.

Most of today's engines have a direct drive to the prop. The geared drive is theoretically more efficient, since it allows the prop to turn more slowly, but it is more expensive to build and can be more maintenance-prone, especially when flown by a pilot with a heavy hand on the throttle.

Now let's talk about a few operating procedures that are often misunderstood. Some of these misconceptions can cost you thousands of dollars over the years.

Use the Throttle Wisely. The first subject has to do with your selection of power settings. After takeoff, many pilots are in a hurry to reduce power because they think this will conserve their engine. This is not true. Your aircraft engine is built for high-speed operation, and you are not doing it any good by lugging along at low power.

Engine failures are rare, but when they do occur, a high percentage of them take place shortly after takeoff. The safest procedure is to maintain full power on climbout until you have gained sufficient altitude to reach an emergency landing site in the event of a failure.

Also, you won't harm your engine by cruising regularly at 75 percent power; the major reason for a lower power setting is fuel economy.

Most pilots know better than to make a sudden large power reduction for descent; the resulting thermal shock could be very hazardous to the engine's health. But what do you do if you're on an IFR flight plan and ATC keeps you up there until you're practically at the airport? What you do is plan ahead, request lower altitudes well in advance, and be a little persistent if necessary. Using your landing gear and flaps for dive brakes (within their prescribed extension speeds, of course) will also help you maintain the RPMs to keep the engine warm.

The Mystery of Mixture. One of the most important controls in the cockpit of your plane is the mixture control—and its proper use may be the least understood of all the knobs and levers you manipulate in flight. Consider the fact that over the years even the engine manufacturers have changed their positions on recommended procedures for mixture control.

The engine is not operating properly if it is running too lean or too rich. The best way to determine the setting right for a given set of power, altitude, and temperature conditions is through the correct use of the EGT (exhaust gas temperature) gauge.

Another important benefit of the EGT gauge is that it can help you to detect and possibly identify certain kinds of engine malfunctions. I can attest to this with some authority. One night some years ago, the EGT gauge in my Mooney Super 21 gave me the first indication of what turned out to be a rather dramatic inflight malfunction: a total loss of oil, leading to a thrown rod. The early warning didn't save my engine, but that was because I elected to keep it running long enough to reach an airport.

In some cases, failure analysis can be made with even greater certainty when the trends of both the EGT and CHT (cylinder head temperature) are observed.

As a pilot, you can exert a tremendous amount of influence over the efficiency and health of your engine through a good understanding of mixture control. At this point I'd like to acknowledge that much of the industry's information on that subject has been generated over a 20-year period by Al Hundere and his company, Alcor, who truly pioneered the field.

What you are seeking to accomplish for maximum efficiency is the complete combustion of both fuel and air in each cylinder of the engine. (This goal cannot be *totally* realized, because neither the carburetor nor even the more efficient fuel injector can meter the identical amount of fuel to each cylinder—but you want to get as close as possible.)

The most effective way to measure the fuel/air ratio is by measuring the temperature of the product of the combustion, the exhaust gas.

When the mixture is rich, the exhaust gas temperature is relatively cool because of the excess fuel. As you lean the mixture, the EGT rises because the excess fuel is reduced and the combustion is more complete. At peak EGT, there is optimum combustion of the fuel/air mixture, and, therefore, optimum efficiency. Leaning past peak will cool the EGT with excess air. If you continue to lean past peak, at some point the cylinder will misfire, due to lack of adequate fuel in the mixture.

Rich misfire can occur at a fuel/air ratio of 13–14 lb of fuel per 100 lb of air. Lean misfire will take place at 4.5–5.5 lb. For best power, you are looking for 8 lb, and for best economy, 6–6.3 lb.

The cruise performance charts in some aircraft operation manuals offer you a choice of best power or best economy mixture settings. The more expensive and relatively inefficient choice is best power.

Operating at the best power setting, which is about 100° rich of peak, will enable you to fly only about 2–4 kn faster than you would at best economy, which is usually peak EGT. However, you'll pay dearly for those few knots with 15–20 percent higher fuel consumption—and that, of course, also reduces your maximum range and/or payload.

Some pilots think by operating well rich of peak they are being kind to their engines. This is not the case. An over-rich mixture can lead to fouled spark plugs and engine deposits. At the other extreme, an overly lean mixture can create an oxygen-rich atmosphere in the cylinders conducive to valve erosion, in addition to the risk of detonation and preignition.

In the past, the conventional wisdom called for cruising at 50° or more on the rich side of peak EGT, and many pilots still hew to this tradition. However, for many engines, Lycoming allows operation at peak EGT for settings of 75 percent power or below, while Continental now endorses peak EGT operation when at 65 percent power or below. The Continental IO-550 engine, introduced with the 1984 A36 Bonanza and 58 Baron, may be operated at normal cruise settings at 20° on the *lean* side of peak.

Of course, you should determine the approved procedures for your particular aircraft, and carefully monitor all temperature gauges, including oil temp, EGT, CHT, and, if turbocharged, TIT (turbine inlet temperature).

For takeoff, climb, and other operations above 75 percent power, leaning is not normally recommended. Bear in mind, however, that at a high-density airport, you might not be able to achieve 75 percent power in a normally aspirated engine. In that case, you may need to lean the engine in order to obtain adequate takeoff power.

There are definite risks involved if you lean while climbing at *more* than 75 percent power. Here are some words on that subject by Ron Roberts, president of Electronics International, a manufacturer of EGT and CHT gauges:

> If you're above 75 percent power, you've got to have enough fuel to cool the charge. And beyond just running a cool charge, you want to prevent detonation, because with detonation, preignition is just around the corner. All it takes is one time, and it'll burn a hole right in the piston or cylinder head. Your combustion temperatures will go from a norm of around 4000 degrees right on up to 6000, and that will literally melt a hole in the piston dome.

Is Your Time Worth $266.67 an Hour? I mentioned that proper leaning procedures will save you money. How much money are we talking about? Well, Alcor ran a test, operating an O-470 engine at 65 percent power at 10,000 ft, and here's what they found:

With the mixture set at peak EGT, the plane trued out at 165 mph and burned 11 gph. At $2/gal, the cost of fuel over 1000 hr of this type of operation would total $22,000. (It does add up, doesn't it?) Enriching the mixture to 100° rich of peak produced a speed increase of 2 mph and a fuel consumption increase of 2.2 gph. That brings the fuel tab up to $26,000, with the shorter time enroute factored in. So at the rich setting, you're paying an extra $4,000 to save 15 hr flight time over the 1000 hr of operation. It's up to you to decide whether or not your time, in this particular example, is worth $266.67 an hour.

Since a single-probe unit will indicate the EGT of only one cylinder, good operating practice calls for the installation of the probe on the exhaust stack of the leanest cylinder. Remember, there is no perfect metering system that can provide identical fuel/air mixtures to each cylinder. Here's where things get a little complicated.

There is no cylinder that is consistently the leanest one. Depending on power setting, altitude, and other conditions, one cylinder may be leanest at one point, and another cylinder may assume that role at another time. So, as Alcor states in its literature, "Because the leanest cylinder of an aircraft engine may change, the conventional single-probe EGT meter has one serious shortcoming—there is no guarantee that the probe is in the leanest cylinder."

That shortcoming can well discourage you from operating at peak EGT as shown on the meter, because another cylinder may then be functioning on the lean side of peak.

The answer is to be able to read the EGT of *each* cylinder by mounting a separate probe on each cylinder's exhaust stack. There are several units on the market that enable you to do just that. Some of the models display the EGT of each cylinder, one at a time, while the more sophisticated units show the readings of all cylinders simultaneously.

The latter devices not only help identify the leanest cylinder, they also indicate the spread, or EGT variation, from one cylinder to another. Alcor claims with this information you can reduce the spread, and thus increase efficiency, by "fine-tuning" the engine in flight through the judicious use of carburetor heat, throttle, or alternate air.

One thing to bear in mind: the leanest cylinder is the one that peaks first, but it is not necessarily the hottest. Therefore, if you're using the type of EGT instrument that looks at only one cylinder at a time, you have to go through a rather complicated procedure to identify the leanest cylinder.

The procedure works like this: pick the cylinder you think is leanest. If you don't have a good basis for guesstimating, start with the hottest one. Lean that cylinder—let's assume it's number 2—to peak EGT. Then go to number 1 and enrich it. If the EGT needle goes

ENGINES AND FUEL

down, that cylinder is not the leanest, so lean it back to where it was. Go through the same procedure with each of the other cylinders. If the EGT needle goes *up* on one cylinder—voilà!, that one is leaner than the others, because the gauge is telling you it was on the lean side of peak. All clear? If not, picture what happens when you enrich a cylinder that's rich of peak. The needle goes down, right? Conversely, what happens when you enrich a cylinder that's on the lean side of peak? That's right, it goes up.

Whichever way you go, you should have good instrumentation for monitoring EGT and CHT—preferably on each cylinder. This equipment should easily pay for itself in fuel and maintenance savings and could save you even more in the event of an engine malfunction.

The Challenge of the Fuel-Injected Hot Start. Every pilot of a fuel-injected engine has his favorite ritual for attempting to start a hot engine. Some even follow the steps recommended in the aircraft information manual—not a bad thought. Certain procedures work better than others, while some routines don't do anything more than raise the temperature of the pilot.

It's a good idea to understand why the engine is so recalcitrant in the first place. The difficulty is the engine's heat accumulates under the cowling when you shut it down, and this can cause the fuel in the lines to vaporize. When that happens, the engine-driven pump cannot move the vaporized fuel in the quantity needed to support combustion. Here is Continental's own recipe:

1. Retard the mixture control to the idle cutoff position.
2. Open the throttle fully.
3. Operate the auxiliary pump in the high position for approximately 20 sec.

This procedure will pump fuel through the heat-soaked lines and purge them, without allowing the fuel to reach the cylinders. At the same time, the lines will be cooled by the fuel, discouraging further vaporization.

After 20 sec have passed, turn off the aux pump, bring the mixture control to full rich, crack the throttle, yell "Clear!" and crank away. If the engine starts, thank Continental. If it doesn't, don't blame me!

Change the Oil. Regular oil changes are one of the cheapest and most effective forms of maintenance you can bestow upon your engine. The oil performs many functions, and takes a lot of abuse in the process. In addition to lubricating, the oil cools the rapidly-moving parts, cushions against shock, inhibits corrosion, and helps keep the engine clean. An efficient oil filter will collect solid particles but not acids and other liquid contaminants that accumulate in the oil, so change it regularly. If you like to change the oil yourself, that's okay, but have it done every so often by an A&P mechanic; he might spot some impending problem you didn't notice.

You might also find your friendly mechanic advising the installation of a new set of spark plugs, even though the old ones are still firing. Before deciding to change mechanics, consider changing the plugs. You see, the electrodes might be doing their job, but the insulation may have become worn to the point where they are no longer efficient at conducting heat away from those electrodes and into the cylinder heads. This can result in preignition, and you don't want that.

Exercise the Engine. Most of us don't get to fly our planes as often as we'd like, and everybody knows it doesn't do an aircraft engine any good to sit idle for a long period of time. So some pilots pop out to the airport ("Honey, I'm going to the Seven-Eleven to get milk for the baby.") and run up the engine for 10 or 15 min. That'll keep the old rust out, yes?

That'll keep the old rust out, no. You're better off staying at home and cleaning out the garage as you promised. As you probably know, condensation occurs each time you start up and shut down your engine. But a short runup will not get the engine hot enough to dry out the condensation. Instead, you'll simply be creating more of the stuff. And protracted ground runups can be harmful to the engine, since they do not permit proper cooling.

The answer is to have a little fun and go fly. A half hour or more at cruise power will fight the oxidation process in the engine, and it will also keep the moving parts of the airframe limber. This prescription should be followed at least twice a month, or better yet, once a week. It'll do *you* good, too.

Engine Departures. As I mentioned at the beginning of this chapter, the current-production piston-engine aircraft are powered by Continental or Lycoming engines that haven't changed much in the past twenty years. But there are some new developments taking place and, predictably, they are coming mostly from outsiders who are looking for a piece of the action.

For example, there is the Thunder water-cooled engine. Liquid is a more efficient coolant than air, and air-cooled engines use a lot of fuel just to keep operating temperatures within limits. The major problem with liquid-cooled engines is weight—and, of course, there's the added complexity and maintenance of the system. The Thunder engine, rated at 700 hp, is presently being tested in the left nacelle of a turboprop Aero Commander.

The Wankel rotary engine has shown promise for use in aircraft. It is relatively light, operates smoothly, and has a low frontal area. Early automotive models developed problems because steel bolts were used on the aluminum block, and the two metals expanded and contracted at different rates, causing leakage. These difficulties have since been overcome.

Some years ago, Curtiss-Wright, with the help of NASA funding, tested a Wankel gasoline engine in a Q-star experimental quiet plane, a Cessna Cardinal, and a Hughes helicopter. Later, they went on to a Wankel diesel engine, which has the potential of offering excellent fuel economy. It might be worthwhile at this point to explain why diesel engines are more efficient than gasoline engines.

A typical gasoline engine burns fuel only to its stoichiometric ratio, which is about 16 parts of air to 1 part of gas. As you open the throttle to increase engine speed, you are adding increased amounts of the 16:1 air-fuel mixture.

Conversely, the diesel utilizes compression ignition not fired by a spark, but heated to the point that whatever combustible is in the cylinder is going to burn. You increase the speed of a diesel by increasing the amount of fuel, rather than air-fuel mixture. The engine is always operating unthrottled, which is the most efficient form of operation.

The John Deere Corporation has purchased the rights to the Wankel from Curtiss-Wright, and it is reportedly interested in continuing to develop a Wankel aircraft engine.

Let's look at some other developments coming out of the private sector. A company has invented what it calls the Dyna-Cam engine, which is a combination of a rotary and a piston engine. A rotating cam is turned by six double-headed pistons, which form 12 firing chambers in two separate engine heads. According to the manufacturer, the pilot can switch off either end of the engine, thus cutting its power and fuel consumption in half. The prototype is rated at 210 hp, and the maker says it has been certified by the FAA.

Another company is experimentally fabricating many engine parts out of fiberglass instead of steel and is getting a considerable weight reduction in the process.

Despite these promising experiments, we must remember changes come about very slowly in general aviation. Both the engine and airframe manufacturers are very conservative, preferring to stay with known quantities. Undoubtedly, this is due in large measure to a fear of increasing their exposure to warranty- and product-liability problems. So we probably won't see many startling innovations under the cowlings of our airplanes for quite a while.

What About Autogas? As the lineperson tops your tanks and hands you a whopping bill for avgas, you might find yourself strongly tempted to lug in some cans of auto fuel from the corner filling station and save, say, 65 cents a gallon. Just think: if your bird guzzles 10 gph, that's a saving of $6.50 an hour, or $9,750 during the TBO life of a 1500-hr engine!

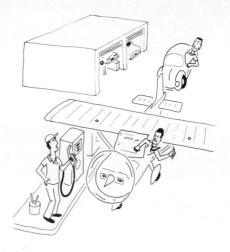

For years, quite a few aircraft owners have done just that — illegally, and, in the official opinion of the FAA and the engine manufacturers, dangerously.

Now there's a breakthrough. Thanks to the pioneering of the Experimental Aircraft Association Aviation Foundation, the use of autogas (sometimes informally called mogas) has been approved by the FAA for a number of aircraft models. This represents a major change in the position of the FAA, which previously had not even considered accepting an application for permission to use mogas.

Why had the FAA, along with the engine manufacturers, prohibited the use of mogas for so many years? The answer can be found in the FAA Advisory Circular 91-33, issued in October, 1971. The circular's main purpose was to deal with the use of alternate fuels at a time when it looked as if 80/87-octane avgas might disappear from the marketplace.

Among other things, AC #91-33 said: "*Do not use automotive gasolines,* even if of equivalent octane or better, as a substitute for aviation gasoline." The reasons presented were as follows:

ENGINES AND FUEL

1. Automotive fuels have a wider distillation range than aircraft fuels and this promotes poor distribution of the high anti-knock components of the fuel.
2. Automotive fuels are more volatile and have vapor pressure which can lead to vapor lock. Automobile gasoline has a vapor pressure of 8 to 14 pounds, varying with different geographical areas and seasons, whereas aviation gasoline has a vapor pressure of 5 to 7 pounds. The low vapor pressure in the aviation gasoline is insurance against vapor lock caused by altitude or heat.
3. Tetraethyl lead in automotive fuels contains an excess of chlorine and bromine, whereas aviation fuels contain only the chemically correct amount of bromine. The chlorine is very corrosive and under severe conditions can lead to exhaust valve failures.
4. Automotive fuels are less stable and can form gum deposits. Gum deposits can result in valve sticking, poor distribution, and other malfunctioning.

As I indicated earlier, many pilots have been literally flying in the face of these prohibitions and warnings. In a survey conducted by EAA's (Experimental Aircraft Association) *Sport Aviation* magazine, 266 aircraft owners reported using a total of nearly a half-million gallons of mogas in standard certificated aircraft ranging from Cubs to Bonanzas. Although these pilots claimed to have had no operational problems, they certainly risked such other difficulties as citations for violating FARs (Federal Aviation regulations), loss of engine warranty, and denial of insurance coverage.

Thanks to a testing program running several years, and painstakingly documented, the EAA has convinced the FAA to reverse its dogmatic position. The first plane to undergo EAA's certification program for mogas was a Cessna 150. The aircraft accumulated 730 flight hours of recorded engineering data through all four seasons of each year. Included were 1200 touch-and-goes, detonation tests with 15 brands of fuel, and vapor-lock tests at altitudes as high as 12,500 ft. One of the tests consisted of using a batch of mogas that had been formulated to 16 in. of Reid Vapor Pressure—more than the maximum RVP distributed to gas stations—and preheating it before flight to produce a "worst case" situation.

The first STCs were sought for aircraft whose engines burn 80-octane fuel. That's because many EAA members own homebuilt or older production aircraft with engines designed for 80/87-octane avgas, which is hard to get in many areas and is in danger of being phased out completely. These engines tend to suffer such maladies as exhaust valve and stem head erosion, intake valve burning, and plug fouling when given a steady diet of 100-octane avgas. Autogas has an octane range that corresponds to 80/87-avgas.

As of this writing, the EAA has obtained mogas STCs for about

200 aircraft models, and has distributed STCs to well over 5000 airplane owners. Another source is Petersen Aviation, a Minden, Nebraska, aerospraying company that obtained STCs for their ag-planes and went on to get approval for a variety of other aircraft. Petersen's STCs permit the use of either regular or unleaded auto gas, whereas the EAA STCs are for unleaded gas only.

There are no operating limitations per se, except the STCs are not approved for Part 135 commercial operations. In order to be legal, you must pay a fee to the holder of the original STC. At present, both EAA and Petersen charge $.50 per hp; EAA adds a surcharge of $15.00 to nonmembers. In return for the fee, you receive the necessary documentation, instructions, and fuel cap placards.

The number of FBOs (fixed base operators) offering autogas is small but growing. *Sport Aviation* periodically publishes lists of these FBOs. When you visit one of these dealers, you'll probably find he has to charge more than the corner gas station, because his cost of doing business is higher and his volume is lower. I would encourage you to deal with any FBO that depends on your business to survive, even if you have to pay a few cents more per gallon. After all, you need the services of the FBO, and it makes good sense to support his operation.

If you find it necessary to truck your gas in from that corner filling station, you'd be well advised to pick a name-brand dealer rather than an off-brand discounter, because some off-brand distributors reputedly dilute their gas with cheaper alcohol-based fillers, which can harm hoses, gaskets, and other components. Also, the fuel should conform to a measurement standard known as ASTM (American Society for Testing Materials) Specification D-439. Finally, it's good practice to use a chamois filter when transferring fuel, as a precaution against contamination.

The cost savings of mogas over avgas are tempting indeed. But before you yield to temptation, be sure to touch all the bases.

5. Propellers

Fixed pitch vs constant speed...Are three blades better than two?...Why are Q-tip props bent backwards?... What's the best exercise for a prop?...Composites are coming.

(Keith Connes)

THE TYPE AND DESIGN of the propeller on your plane will have a definite bearing on its performance, cost of operation, and noise level. On certain types of aircraft, you have a few propeller choices, but in all cases you are limited to whatever has been certified for use with each airplane model.

How efficiently, how safely, and for how long your prop does its job depends a great deal on how well you follow some relatively simple procedures. We'll get into those in just a bit.

The Propeller Is a Very Complex Airfoil. The propeller is, of course, a vertical airfoil whose job is to convert the powerplant's energy into thrust. Its airfoil shape is more complex than that of a wing, because every inch of its span, from root to tip, moves through the air at a progressively higher speed; therefore, its shape must vary progressively from root to tip.

Much of the current prop technology is based on research conducted during World War II. However, the growing interest in energy conservation and noise-level reduction has helped to spur the use of new designs and materials.

Which Pitch? A relatively flat-blade angle, or pitch, allows the engine to rev-up higher for takeoff and climb performance, while a deeper pitch takes a bigger bite out of the air, providing greater thrust at lower engine RPMs. Adjusting the blade angle on a movable-pitch prop is akin to shifting gears in a car.

Generally speaking, the aircraft manufacturers provide fixed-pitch props on planes of 160 hp and under and go to the heavier, more expensive controllable props on the higher-performance aircraft. Here's a rundown on the different propeller types:

FIXED PITCH. This is the simplest type, with no moving parts. Its advantages are low initial cost, low maintenance, and light weight. In one respect, it makes for easier operation, since there's no prop control to fiddle with. However, because it has no governor, you have to monitor RPMs more carefully, especially in a descent. The main disadvantage is that a single fixed-blade angle is inevitably a compromise of the various pitches best suited for takeoff, climb, and cruise conditions.

Many planes using fixed-pitch propellers can be equipped with a choice of standard, climb, or cruise props. As its name implies, a climb prop gives better climb performance than a cruise prop because its blade angle is flatter, but this comes at the sacrifice of a few knots of cruise speed. The tendency today is to go to a standard middle-of-the-road prop designed to achieve a good balance in overall performance.

CONSTANT SPEED. This is a controllable-pitch prop with a governor that maintains the desired RPMs. The pilot selects the blade angle he wants by means of a control knob, thereby activating a hydraulic system that uses engine oil to exert pressure on a piston. The piston produces a force that is opposed by the centrifugal moment of the blades, as well as by a spring.

FULL-FEATHERING. This system is used on multiengine aircraft. If there is an engine stoppage, it is necessary to feather the prop—to turn its blades so they're nearly parallel to the windstream. This stops the prop from windmilling and reduces prop drag to a minimum.

A full-feathering system is basically a constant-speed system with counterweights that put the prop into feather position when oil flows out of the prop piston. This occurs when the pilot puts the prop control into the feather position, or when oil pressure is lost due to engine stoppage or other causes. Some twins are equipped with accumulators that unfeather the propeller when the prop control is moved forward for an easier engine restart.

REVERSIBLE. This system enables the pilot to put the prop into a "beta" mode, reversing the blade angle to bring the plane to a quick stop, or even move it backwards. While the cost of such a system is usually justifiable only for the heavier turboprop planes, the reversible feature can also be very useful in seaplane operations. Hartzell provided a reversible prop for the Seabee amphibian in the 1950s and is making one now for the Lake.

OTHER SYSTEMS. The propellers just described are the ones most commonly found on today's aircraft. Some of the older planes, such as the Navions, are equipped with props that are fully controllable, but are not constant speed, although many have been modified with governors.

Of an even earlier vintage was the Koppers Aeromatic prop, which was designed to adjust its pitch automatically to the prevailing flight conditions. However, it was cumbersome and did not live up to expectations.

A number of ag-planes have ground-adjustable props or simple two-position props that can be set in flight.

Two Blades or Three? Some of the current aircraft are available with a choice of a standard two-blade or optional three-blade prop. Since the three-blade option can cost an additional $1000 or more per engine, and weigh 18-20 lb extra, it would be worthwhile to know what you're really getting for the investment.

A three-blade prop buys you better ground clearance, less noise, and less vibration. The improved ground clearance is achieved because a three-blade prop can do its job with less diameter than its two-blade counterpart. As Chuck Husick, former head of Cessna's McCauley Propeller Division, said, "Ground clearance may be of no importance to you until you run over a bump that's as high as your clearance. Then it becomes very important."

The shorter blades also have slower tip speeds, which cuts down the noise. This is a high-priority item now, due to recent environmental regulations, and as a result, some new aircraft have three-blade props as standard equipment. The propellers on turboprops usually have four blades or more, and NASA has developed one with ten blades whose performance was evaluated at speeds as high as Mach 0.8.

With shorter blades to work with, the designer of a twin can mount the engines closer to the fuselage, reducing the effect of adverse yaw in an engine-out situation. However, this benefit can be accompanied by increased cabin noise.

Then there's the matter of "thump," caused by the propeller slipstream beating against the airplane. A three-blade prop creates more thumps, but each thump has less impact.

Does a three-blade prop improve a plane's climb or cruise performance? Not noticeably. But it looks sexier, and that's a compelling reason for many owners to go for the option.

Some planes benefit more than others by going the three-blade route. A notable example is the Piper Turbo Arrow. Its three-blade prop option reduces vibration considerably, due in part to an accompanying change in the engine-mount system. The Seneca also seems to gain from the option.

Beech provides three-blade props as standard equipment on its turbocharged Bonanzas and Barons. It has been my experience that the benefit of the optional three-blade prop on a normally aspirated Bonanza is strictly cosmetic.

Cessna makes the three-blade props optional on Skylanes, but standard on the larger aircraft.

Mooney does not offer a three-blade prop on its 201 or 231, believing it provides no meaningful advantage. Export Mooneys are available in Germany with a Hoffman three-blade wood and plastic prop, but Mooney believes its performance is inferior to that of their standard props. Mooney's pressurized 301 is expected to have a three-blade prop as standard equipment; the company feels the latest designs are worthwhile, particularly in composites.

What about Q-tips? Perhaps the most widely publicized of the newer-design props is the Q-tip, whose ends are bent aft. The Q-tip prop is manufactured by Hartzell and is presently offered as an option on the Mooney 201 and 231, and as standard equipment on Piper's Cheyenne III. There are also STCs by mod shops for the Bonanza, Baron, Aerostar, and MU-2.

The purpose of this design is to reduce noise and improve prop clearance without sacrificing performance. There is some question about how effectively the Q-tip meets these criteria. McCauley tested the design and told me they found no significant improvement, and the tip is not repairable. When I discussed this with W. B. Harlamert, vice president of engineering at Hartzell, he agreed the tip cannot be repaired, and went on to say that the Q-tip is a special-purpose prop and is not designed to replace those conventional props that are providing optimum performance.

Designer Roy Lopresti states that the Q-tip is bent the wrong way — towards the back — which is actually the "bottom," or flat side, of the airfoil. Citing the winglet as an example, Roy maintains that if you're going to bend the tip, then bend it forward, toward the "top" of the airfoil. And that's exactly what NASA is experimenting with. It's called, not surprisingly, a proplet.

How to Care For Your Prop. Paying a little attention to your prop every time you fly can save a lot of trouble later on. Here's what to do:

AVOID STONES. Every propeller expert I talked to emphasized that one of the best preventive-maintenance steps is to choose your run-up area with care. Keep away from areas that have gravel, stones, or other debris. This will help prevent prop nicks and possibly fuselage or wing

5.1. Q-tip or proplet?

dents as well. If there's no alternative, and traffic permits, you might make your runup on the runway.

EXAMINE YOUR PROP. It's part of your preflight, and with good reason. Even a small nick can lead to trouble, especially in the outer 18 in. of the prop diameter. Small nicks generally can be filed down, but even these fixes should be made by a qualified mechanic. Larger repairs that may affect the balance, structural integrity, or operation of the prop should be handled by a certified propeller repair station.

Also, examine the spinner for cracks and dents. And if it's a controllable prop, check for excessive play.

EXERCISE YOUR PROP. Cycle your controllable prop two or three times during the runup. This is especially important when the temperature is low, because you want to circulate warm engine oil into the hub for proper operation. And when cycling, move the lever all the way back. This helps prevent sludge build up; it's also a more thorough way of checking the mechanical integrity of the system.

DON'T USE YOUR PROP AS A HANDLE. The propeller manufacturers get very upset at the thought of pilots moving their planes in and out of parking spaces by means of the prop. This can cause the blades to be literally bent out of shape. Many pilots compromise by grasping the prop near the hub, but even this technique is frowned on by the manufacturers.

GIVE YOUR PROP REGULAR OIL WIPES. After each flight it's a good practice to go over the prop with a bit of light oil on a rag. This provides corrosion protection and also gives you a good look at the propeller. Regular waxing is also recommended. Don't hose down the prop, and be especially careful to keep water out of the spinner, where it can interfere with lubrication and cause corrosion.

The Props of Tomorrow. The bulk of R & D (research and development) in airfoil design, including props, has been made by NASA and its predecessor agency, NACA. The latest NASA projects include experiments with curved blades, low-drag faired spinners, and composite materials.

Hartzell is now making a composite blade. It is certified at this writing only for the Casa, a Spanish commuter plane. Its main advantages are light weight and corrosion resistance; the drawback is that a composite blade is currently double the price of its aluminum counterpart. However, excess weight costs money every time an airplane flies, and as improved technology brings down production costs, we'll probably be seeing more of the composites.

One interesting prospect is the aeroelastically tailored propeller, whose nonrigid composite blades twist in the wind, so to speak, to get improved performance at various prop speeds. Experiments thus far have shown increases in efficiency of 5 percent and more on both fixed-pitch and controllable props.

As the new designs come off the test stands and onto the production lines, we can expect to see propellers play an increasing role in making planes quieter and more fuel efficient.

6. Landing Gear

Fixed vs retractable...Preventing gear-up landings...Why choose a taildragger?

THE LANDING GEAR is something to think about very carefully when buying or even renting a plane, because the type of gear on the aircraft will have a large bearing on its handling, performance, safety, and cost.

When considering landing gear, you have two major categories to think about. The first category is fixed versus retractable. The second category is tricycle, or tri-gear, versus taildragger. Most airplanes produced in the past 30 years have tricycle gear, but there are circumstances where a tailwheel airplane can offer definite advantages, and we'll explore that in a bit. Let's start with the question, should your plane have fixed or retractable gear?

The Advantages of Fixed Gear. The major benefit of fixed gear lies in its relative simplicity. It is axiomatic of design that the simpler it is, the less likely it is anything will go wrong. Fixed gear is cheaper to build and, therefore, cheaper to buy. It's less expensive to maintain and to check during 100-hr or annual inspections. Each insurance carrier has its own criteria, but the chances are you'll pay a higher premium for a retractable unless you have 300 hr or more of retractable time, and even then, some companies may charge as much as 30 percent more for hull insurance.

Fixed gear will not take up space in the cabin or baggage area, as some retractable systems do. It's also lighter, and normally will reward you with more useful load than a comparable retractable. It will probably stand up better under rough use, such as hard landings and operations on unimproved strips. Finally, you'll never suffer the embarrassment, expense, inconvenience, and possible danger of a gear-up landing — due either to forgetfulness or a systems failure.

The Advantages of Retractable Gear. Higher cruising speed is the main benefit offered by gear that tucks up out of the airstream. Other performance advantages include a superior rate of climb and a higher service ceiling.

A rather bleak benefit occurs in the event of engine failure over water or rough terrain, when the pilot will want to land with the gear retracted to avoid flipping over.

Then there's the strictly emotional factor: many pilots feel more macho, more professional, if they're driving a retractable. That may not be the most practical of reasons, but for many of us, image has a lot to do with our becoming pilots in the first place. (The fixed-gear enthusiasts have their image too — perhaps akin to that of the sensible, no-nonsense types who bought VW Beetles for two decades.)

Table 6.1 and Table 6.2 compare the relative performance specs of fixed-gear airplanes and their retractable counterparts. The two Saratogas have identical powerplants and airframes, except for the landing gear and accompanying structure. Here's how the specs compare:

TABLE 6.1. Comparison of Turbo Fixed-Gear vs Retractable-Gear

	Turbo Saratoga	Turbo Saratoga SP	Turbo Skylane	Turbo Skylane RG
Base price (1984):	$107,560	$131,180	$87,600	$106,500
Horsepower:	300	300	235	235
Std useful load (lb):	1,653	1,621	1,387	1,315
Max cruise (kn):	150	159	158	173
Max range (NM):	960	963	920	1,030
Rate of climb (fpm):	990	1,010	965	1,040
Service ceiling (ft):	14,100	16,700	20,000	20,000

As you can see, the performance gains of the Saratoga SP retractable over its fixed-gear sibling are not very impressive. If it's cruise you're looking for, 9 kn comes rather expensively at $23,620, plus added operating costs. And you lose 32.5 lb in hauling capacity.

The performance differences are more pronounced between the two Skylanes, but you give up 72 lb of useful load in the RG — plus some baggage space, to make room for the main gear.

Now let's compare the normally aspirated fixed-gear and retractable Skylanes, bearing in mind the engines are slightly different.

LANDING GEAR

TABLE 6.2. Comparison of Normally Aspirated Fixed-Gear vs Retractable-Gear

	Skylane	Skylane RG
Base price (1984):	$72,750	$95,800
Horsepower:	230	235
Std useful load (lb):	1,390	1,360
Max cruise (kn):	142	156
Max range (NM):	1,025	1,135
Rate of climb (fpm):	865	1,140
Service ceiling (ft):	14,900	18,000 w/egt

Again there are performance differences that are significant, but they come at a fairly stiff premium in price.

What Makes the Gear Come Up? Suppose you've decided to go the retractable route. Whether you plan to buy or rent, it's smart to know something about the system that makes the gear come up and, hopefully, go down again. In most aircraft, the gear is actuated either by an electric motor or by a hydraulic system that's driven by an electric pump and is sometimes referred to as an electro-hydraulic system. Let's look at the pros and cons of each type of system.

Electric systems are generally faster than hydraulics. A difference of 5-6 sec to get the gear up or down may not seem like much, but in an emergency, that difference could be crucial—especially during the classic sweaty-palms situation where a twin loses an engine on takeoff. The electric systems are also cleaner. On the minus side, they are heavier.

In terms of maintenance, the hydraulic systems can nibble at you with leaky seals, hoses, and the like—but on the other hand, when an electrical system's motor goes, you get a sudden large bite in the pocketbook. Operators whose aircraft experience a lot of gear cycling (for training, commuter service, etc.) prefer hydraulics for their lower overall cost.

There are exceptions to every generality. For example, early Navions have a history of hydraulic problems, due to the fact that the original 1200 or so North American models had a single-piston hydraulic pump that literally pounded the system apart. This was because the plane's hydraulic system called for about 1120 lb of pressure and the pump produced 4000 lb. The offending pump was later replaced with a more subdued model. The Navion, in fact, sports a valve that enables the pilot to turn the system off when airborne and turn it on again prior to lowering the flaps and gear for landing.

Uh-oh! What happens when the gear is supposed to come down and it doesn't? The emergency extension systems have interesting differences. If an all-electric system fails, you may have to get the gear down by cranking a small handle about 50 times, which is good exercise, but not too much fun if you're having other problems at the same time.

Some hydraulic systems—notably those on Piper's Vero Beach retractables—have an emergency setup that allows the gear to free-fall when hydraulic pressure is released.

Neither system is completely fail-safe. If your electric gear motor seizes up, you may not be able to crank the gear down; one exception is found on the Mooney, whose emergency system bypasses the motor. If your hydraulic gear picks up an accumulation of dirt or ice, there's a possibility it might not free-fall completely into the over-center locking position.

Another scenario worth thinking about: should your hydraulic pressure fail in flight, the gear could come down when you don't want it to, instantly reducing your range by about 25 percent—which could present a problem if you're on a long flight over water or unfriendly terrain.

Many of Cessna's gear systems have uplocks to prevent such unwanted extension. In the event of a failure, the gear is extended by means of an emergency hand pump.

Which Planes Have Which Systems?

Traditionally, Beech has gone the all-electric route in their Executive and Corporate aircraft: Bonanza, Baron, Duke, and King Air. Their Aero Center retractables—the Sierra and Duchess—have electro-hydraulic systems.

Sierra and Bonanza buyers can get an option called the Magic Hand, which spookily moves the gear handle to the "down" position when the manifold pressure is less than 18–20 in. Hg and the airspeed falls below 91 kn on the Sierra and 104 kn on the Bonanza. A switch on the panel enables you to deactivate the Magic Hand and make your own decisions. Surprisingly, this safety item, priced modestly (for Beech) at about $1000, is ordered on only about 3 percent of the aircraft and therefore might be discontinued in the future.

Gear retraction time on the Bonanza and Baron is a fast 4½ sec. The Sierra times out at 10 sec.

The Cessna 172RG, 182RG, and 210 have a hydraulic system with an electric pump. From 1974 to 1978, the 210s had gear doors that added to the complexity of their articulated-retraction system. The earlier doors would tend to creep down, causing the gear motor to start cranking and disconcerting the pilot. These and other problems caused Cessna to eliminate the offending doors. Retraction time on a doorless single is 5–7 sec, compared to 10–12 sec for the older 210.

Cessna's 400 series aircraft are also electro-hydraulic, as is the 303

LANDING GEAR

Crusader. The rest of the 300 series planes are electric.

Piper is the only major light plane manufacturer that has no fully electrical gear systems. The Vero Beach retractables—Arrow, Saratoga SP, Seminole, and Seneca—have hydraulic systems actuated by electric pumps.

The Arrow and Saratoga SP have an automatic gear-extension system as standard equipment. It's designed to lower the gear if you forget to, and also to prevent you from retracting the gear prematurely. The system uses a separate pitot mast to sense a combination of prop blast and airspeed. Depending on this combination plus altitude, the gear will automatically extend at speeds ranging from 78 to 103 kn, and it will not retract at speeds below 78 kn.

Some pilots like this automatic feature and others do not. The detractors believe it can cause an accident under certain circumstances. For example, if the engine were to fail, the gear would immediately extend, whereas the pilot might want to keep the gear up for extra gliding distance or a rough-terrain landing. Conversely, he might want to retract the gear at speeds below 78 kn for obstacle clearance. While there is an override lever that can defeat the system, the pilots who oppose the device feel that it's one more complication to deal with in an emergency. They may change their minds if they ever have an unintentional gear-up landing.

The now-defunct Aztec had an all-hydraulic system with engine-driven pumps. Prior to 1979, only the left engine had a pump. If that engine failed on takeoff, the pilot had to pump the gear up manually, while attending to one or two other details. Just before takeoff in one of these vintage Aztecs, many a pilot would instruct his right-seat passenger, "If I yell 'pump,' you pump this handle like crazy while I fly the airplane!" This was not reassuring to first-time flyers, and Piper eventually added a second pump.

The Aerostar has an engine-driven pump on the right engine and an optional auxiliary hydraulic system operated by an electric pump. Considering that a twin is highly accident-prone if an engine fails on takeoff and the gear cannot be retracted quickly, it is incredible to me that this highly necessary backup system is on the options list, especially at a steep enough price (over $2000) to discourage some customers from ordering it. Since I am aware of at least one Aerostar that went in under exactly that circumstance, I advise anyone who buys one to bite the bullet and order the option.

The Navajo has engine pumps on both sides and an electric pump for the gear doors.

The Mooney 201 and 231 have electric systems. Early Mooneys had the famous (infamous to the uninitiated) Johnson-bar manual gear system. After takeoff, the pilot unlatched the bar, which was in the vertical position, took a deep breath and swung the thing backwards to the floor, hopefully without crushing his fingers in the process. If his follow-through was faulty, the lever wouldn't latch down and he'd have to start all over again while the plane staggered erratically into the sky and the passengers either snickered or fainted. For gear extension, the procedure was reversed. Actually the system had one advantage: it was virtually impossible to forget to lower the gear in the pattern. This was not just the flicking of another switch—it was a *production*. (If you think you detect a note of authenticity stemming from personal experience, you are right. I owned one of those Mooneys for three years.)

In 1967, Mooney offered an electric option, and in 1970 the beloved Johnson bar was discontinued.

The Mooney Mite had a similar setup, with an added show-biz effect: if you retarded the throttle with the gear up, a flag would wave in your face. This was ironic, considering the number of times the Mite's gear, after having been dutifully lowered, would collapse on landing.

Then there was the old Bellanca Cruisair: it required 32 rotations of the "bicycle pedal" handle to get the gear up, another 32 to get it down. This tended to discourage full-scale touch-and-goes.

Other Things to Consider

GROUND CLEARANCE. This is important if you'll be flying from unimproved strips. Obviously, the larger the tire size (and the shorter the prop) the better. Some planes have nosewheels that are smaller than the mains; others are identical all around. Wheel fairings add speed but can also jam the wheels with mud or snow.

TURNING RADIUS. Something to take into account if your FBO has a crowded tiedown area. The Grumman-American singles are unique with their free-castering nosewheels—great for precision taxiing but tricky when taking off or pushing the plane back into a parking space.

SHOCK ABSORBERS. Most current planes have the familiar air/oil oleo struts. Some also boast trailing link suspensions, an ideal combination for soft taxiing and landings. Exceptions include the Cessna singles, Piper Tomahawk, Beech Skipper, Grumman Americans, and Champions, which have tubular or spring struts made of steel or fiberglass. Some of these aircraft, especially those with the spring-steel struts, can be quite bouncy if not landed just right.

Other deviants are the Mooney models and the Beech Sundowner and Sierra, which use rubber-biscuit shocks that provide firm arrivals.

Of the above designs, the oleo shocks require the most servicing.

MAX V_{LE}. The aircraft's maximum gear extension speed is something to be considered. The higher the V_{LE}, the more effectively you can use the gear as a speed brake for fast descents without overcooling the engine, or for rapid deceleration in the pattern.

Tailwheel Gear. It used to be called "conventional gear," to distinguish it from the tricycle gear that was unique 40 years ago. Until the advent of the Ercoupe and the Bonanza in the 1940s, most general aviation aircraft had the "third wheel" in the back (Fig. 6.1). Then, in 1950, Piper offered tricycle gear as an option for the Pacer. The following year, a separate nosewheel model was produced. It was called the Tri-Pacer, and it soon eclipsed the Pacer. Cessna introduced its tri-gear 310 twin in 1954, saw the handwriting on the wall, and two years later replaced the 170B with the 172, which became the biggest-selling airplane of all time.

6.1. The Cessna 140 is an example of a "conventional-gear" airplane.
(*Keith Connes*)

Today, taildraggers are definitely the *un*conventional airplanes. However, there are still many applications for which they are superior to their more modern tri-gear counterparts. In fact, if you were simply to observe a taildragger and a tri-gear plane side by side (Fig. 6.2), you'd get a pretty clear picture of the advantages of each.

One obvious difference is that the taildragger sits in a nose-high, climbing attitude, while the tri-gear plane rests in an attitude that approximates level flight. As a result, the taildragger pilot has poorer forward visibility on the ground, and, in fact, it's common practice to taxi many of these planes in a series of S-turns in order to see what lies ahead. However, that same nose-high attitude gives the taildragger

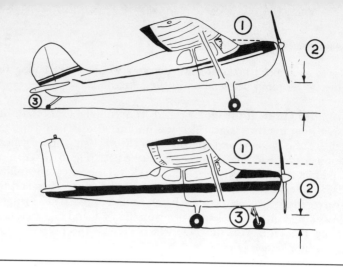

6.2. Some of the pros and cons of taildragger vs. tri-gear are readily apparent. However, this is not the whole story, as you will see.

1. Forward visibility on the ground: the tri-gear wins out.
2. Prop clearance: the taildragger is better here.
3. Weight and drag: smaller is better, so the taildragger wins again.

the benefit of greater prop clearance. This is of minor significance if you use only hard-surfaced airports, but it can be critical if you operate out of rough, unimproved strips.

Another distinction: the tailwheel assembly is small and the nosewheel assembly is large. This tells you that the former has less weight and drag penalties than the latter, which is why all serious aerobatic planes are taildraggers.

However, the tricycle gear plane has a major design advantage that has driven the taildragger into relative oblivion. It is easier to take off and easier to land.

A pilot may be able to maintain heading and altitude with impeccable precision, and his every turn may have the ball nailed to the center, but invariably he judges himself and is judged by others more by the quality of his takeoffs and landings than by any other measurement of piloting skill.

Why is the tri-gear easier to take off and land? Let's start with the takeoff, and we'll assume a crosswind condition. The tri-gear plane starts its roll with its tail in the prop's slipstream, which enhances rudder control. At the same time, the steerable nosewheel helps keep the plane straight.

Conversely, the taildragger starts rolling in a tail-low attitude. The rudder is not much help at that point, nor is the little tailwheel very effective as a steering device. The pilot gets the tail up as quickly as possible, for maximum rudder control, and thus completely loses the services of the tailwheel.

Take a tailwheel plane with a smallish rudder and a large P-factor (the Swift comes readily to mind), throw in a strong left crosswind, and you've got your hands full!

Now let's try a crosswind landing in each type of aircraft. (Fig. 6.3.) The tri-gear plane is landed on the mains, in a nose-high attitude to keep the touchdown speed reasonably low. The pilot can lower the nosewheel as soon as he wants its help for directional control. If he hasn't corrected properly for drift and touches down with some sidewise motion, the geometry of the gear will tend to make the plane want to roll straight.

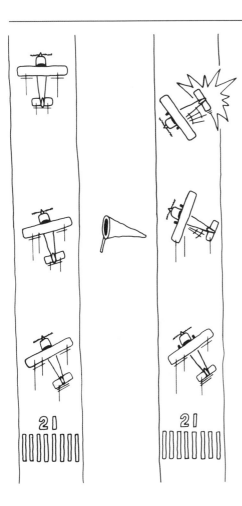

6.3. How forgiving are the tri-gear and taildragger? In a classic landing situation, the pilot of the tri-gear (left) fails to correct enough for the left crosswind and touches down while crabbed left but drifting slightly to the right. With its CG forward of the mains, the plane will tend to straighten out, and its pilot looks like a pro. In the same situation, the pilot of a taildragger (right) makes the same mistake, perhaps without realizing it. The taildragger's CG is aft of the mains, so the tail starts to go, centrifugal force comes into play, and the pilot is introduced to the ground loop.

The taildragger's flight can be terminated with either the conventional three-point landing or a wheel landing. Using the first approach, the plane is landed fairly close to the stall, either on all three wheels at once or possibly with the tailwheel touching first. If the pilot allows the mains to touch first in that attitude, the tail will quickly come down, increasing the angle of attack, and the plane will bounce. Should the pilot then get "behind the stick," he will be rewarded with a humiliating series of bounces.

Suppose the plane has some sideways drift as it touches down. Rather than wanting to straighten out, the taildragger, with its center of gravity behind the main gear, will want to "swap ends" or ground loop.

To help prevent this problem, some Cessna taildraggers were equipped with optional crosswind gear—mainwheels that castered, enabling the plane to weathercock into a crosswind while rolling in angular fashion down the runway, looking somewhat like a strange winged crab. This gear sometimes developed a mind of its own when taxiing, and did not achieve great popularity.

The taildragger pilot may eschew a three-point landing in favor of a wheel landing, wherein the mains are plastered onto the runway while the plane still has considerable flying speed. In a crosswind situation, the pilot lands with a wing down into the wind, setting the upwind mainwheel onto the runway first and then allowing the downwind mainwheel to make contact. Finally, as the plane loses speed, the tail settles. This maneuver requires a decisive manner and an intimate familiarity with the plane's personality.

Am I conveying the message that it takes more piloting skill to handle a taildragger than a comparable tri-gear plane? Yes. And that is one reason some people persist in flying that old-fashioned gear: they like the challenge and the feeling of pride in being "real pilots."

With the exception of the ag-planes, there are very few taildraggers being manufactured today. At this writing, Cessna continues to produce its 185 Skywagon. A few Taylorcrafts are still being turned out. An FBO has bought up the Champion line—Scout, Citabria, and Decathlon—with a view to resuming production. The Super Cub lives on, even though Piper has given up on it. And for those who can handle it, there's the Pitts—magnificent in the air, a bear on the ground.

However, in the used market, there are still many old Cubs, Champs, T-crafts, Luscombes, Pacers, Stinsons, and Cessna 140s, 170s, and 180s around. What's more, there are modifiers who are making Tri-Pacers, Cessna 150/152s, Skyhawks, and Skylanes into taildraggers.

If your flying has been restricted to tricycle-gear planes, you might want to get some dual in a taildragger. You might have to do some searching to find the plane and the instructor—but at the very least, the chances are you'll become a sharper, more confident pilot.

Please forgive the fact that this chapter started out as a coolly objective dissertation on the technical aspects of landing gear and seems to have wound up as a love song to the venerable taildragger. I learned to fly in a taildragger and owned three of them over the years.

Today, most of my flying is done in current factory demonstrators, for the purpose of writing pilot reports on the latest aircraft. But I'll always have a special feeling when I climb into a taildragger.

7. Turbocharging

It's not just for high flyers... Is it worth $10,000 or more?...
Manual, mechanical, and automatic systems... How to combat "coking."

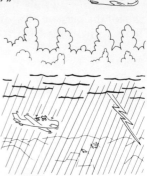

NOT TOO LONG AGO, turbocharging was something found mostly in business twins and a few special-purpose singles. There were a couple of exceptions: in the mid-1960s, Beech offered a turbocharged edition of the V-tail Bonanza; it was a very capable airplane (230-mph cruise versus 203 mph for the normal Bonanza) but did not sell well and was dropped from the line. Then there was the Mooney Mustang — not only turbocharged but pressurized as well; however, a combination of cost overruns and technical problems brought the Mustang to its knees — and the old Mooney company along with it.

Those aircraft were, perhaps, ahead of their time. But times change, and today all of the Big Four offer you a choice of models that are made in both normally aspirated and turbocharged versions. So whether you're buying or renting a plane, it's worthwhile to take a good look at turbocharging and see whether it's for you.

How Turbocharging Works. As you know, the air in our atmosphere becomes less dense with altitude. For the light-plane pilot, this can be both a blessing and a curse, because the airplane moves more easily through the thinner air and thus has the potential to go a lot faster, but the normally aspirated engine can't pump enough of the thin air into its cylinders to maintain full power. As a result, the optimum cruise speed of the average light aircraft is found in the 6000–8000 ft range, above which you run out of throttle and can't maintain 75 percent power. Climb performance, of course, degrades as well.

Enter the turbocharger. It compresses the up-high thin air into the density of the down-low thick air, so at 15,000 ft the engine thinks it's at sea level and performs accordingly. But what about the power to drive the compressor? That's almost the best news of all; it comes virtually as a gift from the engine's exhaust gas, whose energy is put to work instead of merely being piped overboard.

A certain amount of the exhaust gas is routed to a turbine wheel, causing it to spin at speeds as high as 130,000 RPM. The turbine is connected by a shaft to a compressor impeller, which compresses intake air and delivers it to the engine's induction system. To avoid overboosting the engine, the flow of exhaust gas is controlled by a wastegate or, in simpler systems, by a fixed opening. (Fig. 7.1.)

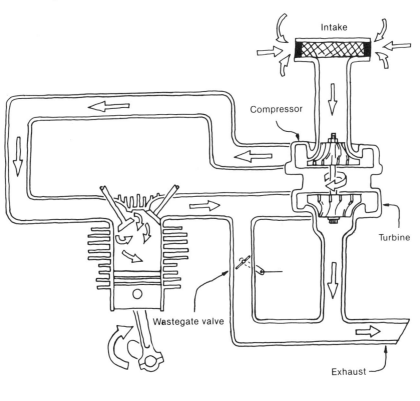

7.1. Turbocharging

In some systems, the wastegates are manually controlled by the pilot, while other systems have automatic controllers. I'll describe these in detail shortly.

All aircraft turbochargers are made by two companies. One is the AiResearch Division of the Garrett Corporation, and the other is Roto-Master. If you're wondering who Roto-Master is and what happened to Rajay, the former purchased the assets of the latter in 1982.

The Advantages of Turbocharging

BETTER CLIMB PERFORMANCE. A normally aspirated engine loses horsepower as it climbs, because it is gulping thinner air with each additional foot of altitude. By contrast, the turbocharged engine in a typical light plane is able to deliver full power at altitudes as high as 20,000 ft.

Most pilots don't climb at full power unless they're hurrying to top weather or mountains. Therefore, they usually don't experience the turbo difference until they go to high altitudes and/or high air temperatures. For example, because of engine-cooling considerations, the Turbo Skylane RG has a sea-level rate of climb *less* than that of the normally aspirated Skylane RG, and on a standard day at cruise climb, both airplanes will arrive at 10,000 ft within one minute of each other.

FASTER SPEEDS AT HIGH ALTITUDES. If you're going to do all your flying at around 6000 ft, don't waste your money on a turbocharged plane, which at that altitude may fly even *slower* than its normally aspirated sibling, due to cooling drag. But once you get up to the 12,000-ft level, the difference becomes noticeable, and if you're willing to fly *really* high, it's like stepping up into another class of airplane. For example, here are the optimum cruise comparisons for three normally aspirated aircraft and their turbocharged siblings:

Mooney 201:	169 kn/6,000 ft
Mooney 231:	192 kn/24,000 ft
Arrow IV:	143 kn/6,000 ft
Turbo Arrow IV:	166 kn/19,000 ft
Skylane:	143 kn/6,000 ft
Turbo Skylane:	159 kn/20,000 ft

Of course, you noticed that the turbocharged models did their best at high altitudes. But you don't have to fly that high to get a speed advantage; all of the turbocharged airplanes have a significant cruise edge at the 12,000-ft level.

HIGHER CEILINGS. This goes beyond the cruise-speed differences noted above. Obviously, the actual *groundspeed* advantage at high altitude can be either enhanced or nullified by the winds aloft. Probably the most important advantage of a plane with a high service ceiling is its ability to top weather. This can often mean the difference between go and no-go. And if icing is encountered and a higher altitude is indicated, it is indeed comforting to have ample climb power at hand.

High-altitude flying is also a good insurance policy for mountainous operations, single engine at night, water crossings, etc.

BETTER HIGH DENSITY ALTITUDE TAKEOFFS. This is another big sales clincher for the turbos. A fully loaded airplane, a strip that's short and/or obstructed, a high field elevation, and a hot day make for a combination that's certain to induce a case of white knuckles. A turbocharged plane is a good remedy, as long as you respect its limitations and your own.

INCREASED MULTIENGINE SAFETY. Turbocharging can provide dramatic improvements in the single-engine performance of twins, especially the light twins that tend to be sluggish on one fan. A case in point: the normally aspirated Seminole has a single-engine service ceiling of 4100 ft, while the Turbo Seminole's single-engine service ceiling is 12,500 ft.

The Disadvantages of Turbocharging

HIGHER COST. The list-price difference between a normally aspirated single-engine airplane and its turbocharged counterpart runs from about $12,000 to more than $20,000. You might think you'll make up a good part of the cost difference in fuel savings, but then again, you might be kidding yourself. One Turbo Arrow owner said to me, "I fly between 9000 and 11,000 ft, and at that altitude I'm 25 kn faster than in a normal Arrow. If you fly a couple of hundred hours a year, mathematically your fuel cost makes it absolutely essential that you buy a turbocharged airplane."

I don't know how he came to that conclusion. According to the Piper information manuals, if he flew 30,000 NM a year, he used approximately 500 gal *more* avgas than it would have taken to fly the same distance at the same altitude in a normally aspirated Arrow. (Incidentally, production of the normally aspirated Arrow was discontinued starting with the 1983 model year.)

Also, the aircraft owner must plan on overhauling his turbocharger system when he overhauls his engine. The cost for an overhauled system on an exchange basis could run $2000–$4000 per engine, depending on the complexity of the system.

MORE PILOT ATTENTION. A turbocharged engine requires more fiddling with the power controls than does a normally aspirated engine. Also, the gauges must be monitored with extra care. More about that later.

THE ANNOYANCE OF OXYGEN. Some pilots won't go up to altitudes that require supplemental oxygen, turbocharging or no—but those who do will have to put up with the discomfort of the mask and the hunt for refills, unless they go first class with pressurization.

Which Turbo System? There are major differences between one system and another, and the type of system that comes with your plane will have a direct bearing on how much work you'll have to do and how much efficiency you'll get out of it.

Here are the systems:

FIXED ORIFICE. This is the least sophisticated of the turbocharger systems. It is sometimes called a fixed-wastegate system; however, that's a misnomer, since there is no wastegate, but merely a bypass located between the turbine inlet and the exhaust outlet. The bypass has an adjustable opening that has been set at the factory to allow full-throttle operation at about 12,000 ft density altitude.

Below that density altitude, do not firewall the throttle on takeoff, or you will exceed the manifold pressure redline. For example, on the Turbo Arrow, MP redline is 41 in. The takeoff technique is to advance the throttle to about 25 in., pause for a second or two while the turbine spools up, then resume advancing the throttle until the MP needle is at the 40-in. mark (it'll drift up another inch). If you were to exceed 41 in., you'd get an amber warning light. There's a relief valve to protect the engine against overboost, but it's best not to rely on that.

Obviously, your takeoff procedure now has an added complication, since you must eyeball the MP gauge (which, on the Arrow, is located inconveniently at the bottom of the panel) while checking your airspeed and keeping the plane straight on the runway.

As you climb, you must add throttle to maintain the desired MP, just as you would with a normally aspirated engine. However, the difference is you will still see 41 in. MP at 12,000 ft density altitude, with all 200 hp at your disposal. And you will be able to maintain 75 percent power right up to the Turbo Arrow's maximum certified ceiling of 20,000 ft.

The fixed-orifice system is used on the Turbo Arrow, Seneca II/III, and Mooney 231. It is a simple, low-cost system that does its job, but does not have the flexibility and overall efficiency of the more sophisticated systems.

MANUAL WASTEGATE. This is the system found on kits that were made by Rajay for the retrofit market. The wastegate is operated by a separate knob on the control panel. For takeoff, the wastegate is kept fully open, and then is gradually closed by the pilot as he begins to run out of manifold pressure during climb.

There are some significant disadvantages to this sytem. For one thing, the pilot must be sure the wastegate is open during takeoff and upon descent from altitude. Since many installations do not provide overboost protection, the application of full throttle at low-density altitudes could damage or completely destroy the engine.

Then, too, a retrofit kit is often bolted onto the aircraft's original-equipment high-compression engine, which is not ideal for turbocharging. The factory installations utilize low-compression engines that can tolerate higher boost, and the complete turbocharging systems are usually packaged, and therefore blessed, by the engine manufacturer.

Earlier turbocharged Aerostars had electrically operated wastegates and used the same high-compression engines found on the normally aspirated models. However, the turbocharged Aerostars now being manufactured by Piper have low-compression engines and automatic controllers.

I do not mean to suggest that you should stay away from retrofit systems. For one thing, some mod shops have STC'd power packages that have modern automatic controllers and should work very well. But if you go into the used-plane market and buy an aircraft with an old-fashioned manual system, you should be aware of its limitations and the need for operational care on your part.

The only turbocharger manufacturer that made systems for retrofit use was Rajay Industries, which at one time provided kits for a variety of aircraft. However, in later years, Rajay reduced the number of kits it offered. The successor company, Roto-Master, is concentrating on the OEM (original equipment manufacturer) business—that is, supplying turbochargers to the engine manufacturers. However, they have licensed a company called Century Aircraft to handle retrofits.

THROTTLE-LINKED WASTEGATE. This system incorporates a mechanical linkage between the throttle control and the wastegate. On some aircraft, the wastegate starts to close as soon as the throttle is advanced, while on others the throttle plate is fully opened before the wastegate begins to close.

As with the fixed-orifice system, the pilot must monitor the MP gauge so as not to exceed redline, and he must continually advance the throttle during climb to maintain the desired power.

The throttle-linked wastegate system is used on the Cessna Turbo Skylane and Turbo Skylane RG, as well as the Rockwell 112TC and the Piper Lance/Saratoga.

AUTOMATIC SYSTEMS. There are four different types of automatic wastegate controllers in general use: the fixed-point controller, variable absolute controller, density controller, and sloped controller. Each works in a somewhat different way, but basically they all serve the purpose of automatically adjusting the wastegate to maintain a given power setting. This eases the pilot's workload, as compared to the systems described previously. For example, with an automatic system, the pilot can firewall the throttle on takeoff and the controller will hold the power at redline. Likewise, whatever power the pilot sets for climb will be maintained with little or no throttle adjustment. And there will be fewer power fluctuations during cruise.

Turbo Flying Tips. If you decide that turbocharging is for you, get a thorough checkout on the system you'll be flying and read the aircraft and engine manuals carefully. Also, if you're going to fly at altitudes requiring supplemental oxygen, get checked out on the use of the equipment. And be sure you understand the FARs and air traffic control procedures that pertain to flight at high altitudes.

Have your MP, temperature, and fuel-pressure gauges checked for accuracy, and monitor them constantly in flight. Remember, the higher you go in a turbo, the higher the engine temperatures become, because the more air is compressed, the hotter it gets. (This problem is avoided with engines equipped with intercoolers.) Also, watch for signs of fuel vaporization, which may occur more readily during the spirited climb of a turbocharged plane. Be prepared to use the boost pump if fuel pressure fluctuates.

Since engine oil plays an important part in the operation of the turbocharging system, it should be given adequate time to warm up. On the first flight of a cold day, bring in the power more gradually than usual. And prior to takeoff in a twin, it's good practice to apply at least 25 percent power with the brakes set, to make certain the engines are developing power evenly.

Allow at least 5 min of low-RPM operation before shutting the engine down. Otherwise a condition known as "coking" can occur.

Here's how that process takes place: The turbine's center housing is lubricated with engine oil, and some of it accumulates on the wall of the housing, which is not up to turbine temperature, but is still quite hot. When the engine is shut down, there is a soakback of heat from the turbine housing to the cooler center housing. This causes the oil to degrade and break down into carbon, which builds up inside the center housing wall. To aggravate matters, the carbon acts as an insulator, thus accelerating the buildup of additional carbon.

Coking cannot be avoided entirely, but it can be minimized by observing a cooling period prior to engine shutdown. Often, the time it takes to taxi from the runway to the parking area will suffice.

The most important maintenance items are clean engine oil and clean air. Remember, a speck of grit can play havoc with a turbine shaft at 90,000 RPM. According to AiResearch, if the turbocharger gets clean oil and clean air, it should run to the TBO of the engine. A turbocharged engine is a little harder on the oil than a normally aspirated engine, because the environment is hotter and there are additional lubricating chores; but normal oil change intervals of 50 hr should still be adequate.

The decision to go turbo depends largely on the kind of flying you do and the kind of performance you want from an airplane. You might try renting before buying, just to get the feel of the different systems. One thing is certain: if you haven't flown a turbo yet, it'll be an exhilarating experience for you.

8. Supplemental Oxygen

You may need it as low as 5000 ft...Built-in and portable systems...New lightweight bottles...Can the cannula replace the mask?...High flight in an altitude chamber.

FEDERAL AVIATION REGULATION Part 91.32 specifies the minimum flight crew must use supplemental oxygen on flights of more than 30-min duration above 12,500 ft and up to 14,000 ft cabin-pressure altitude. Above 14,000 ft, the crew must use oxygen at all times, and above 15,000 ft, each occupant must be provided with supplemental oxygen.

That seems straightforward. "Fine," you say, "if I'm going to pilot a plane above 12.5, I'll have oxygen on board." But you really should know that:

1. You may need supplemental oxygen at *considerably lower* flight levels, particularly at night.
2. There are different types of oxygen systems and different types of masks. It's important to know what you're getting.
3. There's an inexpensive accessory designed to give you the proper amount of oxygen for each flight level.

Let's start with a reminder of why we need supplemental oxygen at higher altitudes. The air we breathe near the earth's surface consists of 21 percent oxygen, which we use to convert the fuel from our food into heat, energy, and life itself. As we ascend in an airplane, the air pressure, or density, decreases. At 10,000 ft it's about two-thirds the density of the air at sea level, and at 18,000 ft it's one-half the sea-level density. But since the percentage of oxygen in that thinner air remains at a constant 21 percent, our lungs are getting less oxygen at that altitude.

Hypoxia—Rapture of the Heights. Lack of sufficient oxygen goes by the medical term *hypoxia*. It is an insidious malady, in that its early symptoms resemble that of mild intoxication; judgment is often impaired to the point where you become euphoric and don't realize what is happening to you. What *is* happening, in all likelihood, is an increasing drowsiness and lack of coordination. As your condition deteriorates, a headache and blurred or tunnel vision probably occur; your heart quickens, your lips and the skin under your fingernails turn blue, and finally, you pass out.

What about the FAA's magic oxygen altitude of 12,500 ft? It's strictly rule of thumb. Just as hypoxia symptoms vary with individuals, so do altitudes at which supplemental oxygen is necessary. Two important factors that can affect your need for oxygen are your age and whether you smoke cigarettes. As you get older, your lungs tend to harden and lose some of their ability to bring oxygen into the bloodstream. If you smoke, you are coating the alveoli, or air sacs, in your lungs with tar—in effect, ageing them. What's more, cigarette smoke is about 4 percent carbon monoxide. According to the FAA, a one-pack-a-day smoker will have the same oxygen requirements at 10,000 ft as a nonsmoker at 14,000 ft.

Night flying brings its own special oxygen requirements, because as the oxygen content of your blood decreases, so does the ability of your eyes to absorb light. Your night vision could be impaired after a prolonged flight at 5000 ft or less. Many professional pilots will breathe oxygen for a few minutes prior to landing. In fact, some foreign airlines require this of their pilots because, even though the aircraft are pressurized, cabin altitudes get up to about 8000 ft.

Built-in vs Portable Systems. With a built-in system, the oxygen bottle is usually stored in a forward or aft baggage compartment, and lines lead to outlets located in the headliner or at the sides of the cabin. When oxygen is needed, occupants of the aircraft plug their masks into the nearest receptacles. The pilot activates the system by pulling a control handle connected to a cable that opens a valve at the bottle. In a portable system, one or two bottles are placed centrally within the main cabin area, and masks are plugged directly into a regulator on the unit.

The advantages of the built-in system are that its capacity can be greater, the bottle doesn't take up any cabin space, and the whole thing is generally neater. On the other side of the coin, the portable system costs less, it can be transferred from plane to plane, getting refills may be easier, and CG problems are less likely.

This last item should command your attention. A large oxygen bottle can weigh more than 30 lb, so you'll have to consider its location and the effect on the aircraft's weight and balance.

All of the major aircraft manufacturers offer built-in oxygen systems on their turbocharged and pressurized models, usually as an extra-cost option. As an example, the built-in system available for the Piper Turbo Saratoga consists of a 64-cu-ft oxygen bottle with remote filler, pilot's oxygen mask with built-in microphone, control-wheel mike button, and five passenger masks. The system weighs 40 lb and costs about $3000.

In addition to getting a factory installation, you can have a built-in system installed in your plane through a dealer or mod shop, assuming the aircraft can carry it. And even a small plane should be able to accommodate some sort of portable system.

Constant Flow. Most light aircraft are equipped with what is called a constant flow system; that is, once the system is activated, it supplies oxygen at a fixed rate, regardless of the rates at which the users are breathing. When planning to fly at high altitudes, bear in mind the need for supplemental oxygen increases as the altitude increases. The following data provide recommended flow rates for constant flow systems, as published by the FAA's Civil Aeromedical Institute:

Altitude (ft)	Oxygen Flow (L/min)
5,000	0.5
10,000	1.0
15,000	1.5
20,000	2.0
25,000	2.5
30,000	3.0

Since 1 cu ft equals 28.32 L, a 22-cu-ft bottle would supply one person with an estimated 5.2 hr of supplemental oxygen if it were set to deliver 2 L/min for flight at 20,000 ft. ($28.32 \times 22 \div 2 = 311.52$ min, or 5.192 hr.) But if you were flying at 10,000 ft, you might need a flow of only 1 L/min and would be wasting oxygen at the previous setting. So although the simplest systems are either on or off, some systems can be adjusted manually for various altitude ranges, and others have automatic altitude-compensating mechanisms.

You should compare flow systems and bottle sizes and check the manufacturers' spec sheets to determine what kind of duration you can expect from each. If you're buying a new plane with a factory-installed system, the option list may not offer much choice. But if a system of another type or capacity would meet your needs better, tell the dealer what you want and see if he can provide it.

Important: when planning a flight at high altitude, don't rely completely on any flow chart or spec sheet for estimating oxygen duration. Allow ample margin for variations in rates of breathing; for example, in a stress situation, your breathing rate might increase. Also, there can be some leakage in the lines and fittings. Monitor your cylinder pressure just as you would any important instrument.

SUPPLEMENTAL OXYGEN

Demand. In addition to the constant flow system, there is the demand system, generally used in the more sophisticated installations. The demand system causes oxygen to flow only when the users are inhaling. This conserves oxygen, but the equipment is more complex and, therefore, more expensive. Some installations combine a demand system for the crew and a constant flow system for the passengers.

Masks. There are two types of mask systems available for general aviation use. Probably the one more commonly used is the rebreather mask. With this, oxygen flows into a rebreather bag, which in turn supplies the air you inhale. Your exhaled air goes back into the bag and is mixed with fresh oxygen; thus you are rebreathing the air you exhaled, along with incoming oxygen and, in some systems, a certain amount of ambient (cockpit) air as well.

The rebreather system has the advantage of recapturing unused oxygen, since you don't use all the oxygen you inhale with each breath. Also, the "recycled" air is moister, warmer, and thus more comfortable than pure oxygen, since part of it comes from your lungs. (Don't worry about taking in the CO_2 you've expelled; the mixture you get is not harmful.) Rebreather equipment is not recommended for use above 25,000 ft.

The second type is the phase-dilution system. This also utilizes a bag attached to the mask, but it's called an accumulator bag, and you do not exhale into it. Oxygen feeds into the bag, and you inhale it; when your inhalation exhausts the oxygen in the bag, a valve in the mask opens and the remainder of your inhalation consists of cockpit air. The benefit of this system is that the initial part of your inhalation, which goes deepest into your lungs, is pure oxygen. It can, however, be drying to the throat over a long period of time. The phase-dilution mask is generally more expensive than a comparable rebreather mask.

Whichever type of mask you choose, get one that fits your face snugly and comfortably. A built-in mike is desirable, but try before you buy. Some work well, but some are virtually unreadable; a lot depends on mike placement, type, and quality.

Ted Nelson's Useful Device. The Nelson Oxygen Flow Meter is an accessory that, for a relatively small sum, can do a big job in getting you exactly the right amount of oxygen for the altitude you're flying.

To understand what a flow meter is, let's start by comparing it with what it replaces on most general aviation systems. The FAA requires all supplemental oxygen systems to be equipped with an indicator showing whether oxygen is flowing to each mask. The usual method is to install a green and red capsule in the breathing tube. When oxygen is flowing, the capsule shows green. What it does *not* show is how *much* oxygen is flowing.

According to Ted Nelson, the standard capsule will show green if as little as ¼ L/min is flowing. So Nelson, a retiree in his seventies, has developed and marketed a meter that replaces the capsule and shows exactly how much oxygen is flowing. But it's more than a meter; it's a control containing a little thumbscrew that enables you to fine-tune the amount of oxygen you're getting. If you're flying at 20,000 ft, turn the valve until the ball in the window floats up to the big 20, and *voilà!*—you'll be getting 2 L/min of oxygen.

At first Nelson tried to market his invention as a safety device, but nobody cared. Now he sells it on the basis of how much money is saved by not wasting oxygen, and he's doing fine. (That little story has an interesting moral.) Nelson makes several models, priced in the $30 to $50 range. He also sells complete oxygen systems, including masks—one of which has a noise-cancelling dynamic mike he claims really works.

Another Way to Breathe. Doctors Sidney White and Bernard Diamond have been marketing something called the EZ/Ox oxygen nasal cannula, which is an adaptation of a medical device (Fig. 8.1). The cannula enables aircraft occupants to breathe supplemental oxygen through the nose instead of the mouth. It consists of a length of plastic tubing formed into a loop. In the center of the loop are two openings, with tips that fit into the nostrils. The ends of the loop are joined into a plug that goes to the oxygen supply. The virtue of the cannula is that it eliminates the discomfort and restrictions of the oxygen mask, allowing the users to talk, eat, and do everything else but smoke.

Until recently, the mask has been the only legal device for breathing oxygen where its use is required by the FARs, but the cannula has now been approved as an alternate method, up to a maximum altitude of 18,000 ft. White and Diamond hope a higher altitude will eventually be permitted. The cannulas are expected to sell for about $7.00 each.

Using and Maintaining Your System. Once you've got the system, make sure it's ready when you are. Include the oxygen system as part of the preflight. Check to make sure there is an adequate supply of oxygen. Turn the system on and take a few sniffs to make sure the oxygen is flowing properly and doesn't smell contaminated. Make sure all hoses and masks are in good shape and accessible.

8.1. The cannula provides supplemental oxygen and lets you talk or eat. (*White Diamond Corp.*)

Do not use all of the oxygen supply before refilling the bottle, or the bottle will have to be purged, which takes time and could possibly be done improperly.

Note the date by which the bottle will have to be hydrostatically tested. Standard-weight cylinders have the DOT rating of 3AA 1800 and must be retested every five years, with no limit on service life. Lightweight cylinders have a Department of Transportation rating of 3HT 1850, must be retested every three years, and are limited to a service life of 24 years of 4380 pressurization cycles, whichever comes first.

Who Makes What? Here are the Big Three of the supplemental oxygen business and a listing of their major portable systems:

SCOTT. Their Executive Mark I has an 11-cu-ft cylinder and supplies oxygen for one or two persons up to 16,500 ft. The Executive Mark II comes in sizes ranging from 11–63 cu ft, supplying one to four persons up to 16,500 ft or, with optional accessories, to 23,000 ft. The Executive Mark III comes in a seat-mounted console case, 22- to 44-cu-ft capacity, supplies one to four persons up to 19,000 ft, or 27,000 ft with accessories, and features automatic altitude compensation. (Fig. 8.2.)

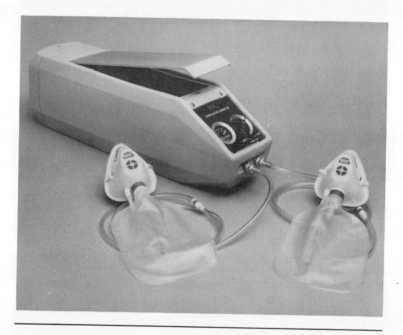

8.2. The Scott Executive Mark III console, a portable supplemental-oxygen system. (*Scott Aviation Products*)

PURITAN-BENNETT. Their system Number 176932 is for one or two persons, with 11-, 15-, or 22-cu-ft cylinders, and features a choice of altitude ranges: 8000–14,000 ft; 14,000–22,000 ft; 22,000–30,000 ft. System Number 176934 serves up to four persons and has the same altitude ranges as the smaller system. Cylinder sizes are 22 and 38 cu ft. The Altitude Traveler comes in a carrying case, serving up to four persons to a maximum altitude of 30,000 ft, using a 22-cu-ft cylinder.
SKY OX. This company, formerly affiliated with Rajay Industries, was sold in 1982, but operation is continuing virtually unchanged. Their portable systems are contained in a pack that can be suspended from a

SUPPLEMENTAL OXYGEN

seat back, as opposed to the configurations that use a console case. The SK-9 model provides oxygen for one or two persons, with a choice of 20,000-ft or 25,000-ft altitudes and 20-, 35-, or 50-cu-ft cylinders. The SK-10 serves up to four persons, with the same altitude and cylinder choices as the SK-9.

What's New? The systems we've discussed use bottles made of steel, a material that has been the standard of the industry. Now, bottles are being offered that are aluminum wrapped with Kevlar® fiber. These bottles are about half the weight of comparable steel bottles. They are also about 50 percent more expensive but could be worth the extra cost when maximum payload is important. (Fig. 8.3.)

8.3. Aluminum and Kevlar oxygen cylinders weigh less but cost more. (*Scott Aviation Products*)

Then there's the solid state system that supplies oxygen via combustion of a chemical "candle." Scott manufactures such a system under the name Aviox. At present, it's being marketed as an individual emergency system for corporate and airline use. The Single-Pak weighs 4.3 lb complete and will supply 4 L/min for 20 min; the Dual-Pak weighs 8.25 lb and supplies twice the amount.

If you're considering a supplemental-oxygen system, do some homework and shop around. Do you want a built-in system or a portable? Rebreather masks or phase dilution? An 11-cu-ft bottle, 117-cu-ft, or something in between? (The bigger the bottle you can carry comfortably, the cheaper the refills, since most FBOs have a minimum service charge that makes small-bottle refills outrageously expensive.)

If you haven't used oxygen and want to see how you like it, you can rent a portable unit from some FBOs, or if that's not available, rent an oxygen-equipped plane and tool around for a while. But whatever you do, get checked out on the system before you use it for real. Also, make sure your passengers know how to use their masks, and check regularly to make sure they're okay back there.

Supplemental oxygen can be a very handy friend. It can help you top the weather, take advantage of tailwinds, and be a sharper flyer day and night. That's a lot of good things to pull out of the air.

Take a Class. The U.S. government offers a one-day physiological training program aimed at understanding and surviving in a flight environment at high and low altitudes. The course includes such topics as supplemental oxygen, stress, hyperventilation, ear pain, and gas expansion.

Also included is a simulated flight in an altitude chamber to 25,000 ft. At that altitude, you remove your mask and experience hypoxia for no longer than 5 min. You'll be given some simple math problems to perform, such as reciprocal headings, to see what effect hypoxia has on your piloting chores. You'll also observe the reactions of other participants.

On the way back down, you will practice pressure breathing. Depending on where you go for the training, you may also be given a rapid decompression from 8000 to 18,000 ft.

What will all this do to your body? According to the Civil Aeromedical Institute, there is some gas expansion taking place in the body at that altitude, but not enough to cause real discomfort.

Who needs this training program? The FAA feels everyone who flies should participate. Among other things, the altitude-chamber flight will give you a true picture of your reaction to hypoxia under safe and controlled conditions. As an added testimonial, most of the

SUPPLEMENTAL OXYGEN

professional pilots who fly for Cessna's Air Transportation Division voluntarily retake this training every two years or so.

The training is available at a variety of military bases, as well as the FAA Aeronautical Center in Oklahoma City and the NASA-Johnson Space Center in Houston. It's free at the latter two facilities, but it costs $20 at the military bases. Pick up an application at your local GADO, or write to the Civil Aeromedical Institute, Airman Education Section, AAC-142, P.O. Box 25082, Oklahoma City, OK 73125.

9. Pressurization

Fly high without the bottle... The new p-singles are coming... Max differential explained... Piloting the pressurization system.

Before

After

IN THE PREVIOUS CHAPTERS, we discussed turbocharging and supplemental oxygen systems, which are closely related to each other. Both systems are designed to enable you to enjoy the benefits of high-altitude flying by circumventing one of Mother Nature's laws, to wit: the higher you go, the less the air pressure.

Since neither reciprocating engines nor human beings function very well at five-digit elevations, we pressurize the air in the cylinders to make the engines go and sniff oxygen to keep ourselves going.

But wait a minute. Surely there's a better way for *us* to go than to sit hooked up to bottles like hospital patients. Of course there is! If we can stuff extra air into the engine, why can't we also stuff extra air into the cabin, and breathe *it* instead of thin air spiked with bottled oxygen?

Well, we can. All we need is a pressurized airplane.

The $65,000 Difference. At this writing, the list price of a Pressurized Centurion was about $65,000 higher than that of a Turbo Centurion. Is that price difference justified?

Well, I know that some prices in this industry are arrived at rather artificially, but I also know that the design and construction of a pressurized plane costs a lot of money.

You cannot simply take a standard airplane and install a pressurization system. The cabin structure would not be strong enough to withstand the necessary pressure, and, in any case, air would leak out as fast as you could pump it in. So even though the P-Centurion looks very much like its unpressurized counterpart, it is a very different airplane. Its construction includes double-row riveting, sealant, back-to-back formers, additional reinforcement members, and heavier-

9.1. The Piper Malibu, a pressurized single-engine plane. (*Piper Aircraft Corp.*)

gauge channels. The single cabin door and emergency exit are stronger and have inflatable seals. The small, round windows, typical of pressurized aircraft, provide a sturdier, more airtight cabin—or, in industry jargon, pressure capsule.

And, of course, a pressurized plane must go through much more involved fatigue testing and other certification procedures.

From 1978 to 1983, Cessna had the pressurized single-engine field all to itself with the P-Centurion. In 1983, however, Piper's Malibu received certification and went into production. Not only is it a pressurized single, but it's a cabin-class plane to boot (Fig. 9.1). (What makes a plane cabin class, as opposed to, say, tourist class, is an airstair door.)

Piper pulled kind of a quarterback sneak on Mooney, which some time ago announced its own pressurized single, the 301, but at this writing, certification has been delayed indefinitely due to a lag in the economy.

Actually, the old Mooney company started the whole thing back in 1966 with the introduction of general aviation's first pressurized single. The Mustang was a five-place airplane powered by a 310-hp

Lycoming engine, whose performance was quite similar to that of today's P-Centurion. The Mustang died from an overdose of fiscal optimism; it was priced around $30,000, and firm orders were taken for about 30 airplanes, but actual production costs turned out to be far in excess of the selling price. You don't stay in business that way, and, in fact, the original Mooney company didn't. There are still approximately 15 Mustangs in license.

Another bright hope that was started over a decade ago and is still struggling to emerge is the Skyrocket by Bellanca Aircraft Engineering (a different company from the Bellanca that made the Vikings, Citabrias, etc.). The prototype, with its sleek aluminum honeycomb and fiberglass construction, has achieved speeds of close to 260 kn, or 300 mph, on 435 hp. The prototype is not pressurized, but production models are slated to be—if the project ever gets sufficient funding.

The other two pressurized singles looming on the horizon are jetprops. One is the Beech Lightning, which utilizes the fuselage of the P58 Baron. Production was originally slated for late 1984 or early 1985, but it has been postponed indefinitely at this writing.

The second contender is the Smith Prop-Jet, a project of Mike Smith, who modifies Bonanzas to make them faster and safer. Like the Skyrocket, the future of the Smith Prop-Jet is dependent on the successful raising of considerable capital.

Then there's a very interesting plane that's not a single, but neither is it a twin in the usual sense. I refer to the Lear Fan, the last aviation project to be initiated by Bill Lear before his death in 1978. The Lear Fan is powered by two turbine engines linked to a single propeller at the rear of the plane. This aircraft, backed largely by the British government for intended production in Northern Ireland, has suffered massive cost overruns, and its future is in jeopardy.

Keeping the Pressure Up. To understand the value of pressurization and the need for proper operation of the equipment, let's take a moment to explore the physiological effects wrought on our bodies by changes in altitude.

The air surrounding us is affected by the earth's gravitational pull. This pull is expressed in weight, or pressure, per square inch. At sea level under standard atmospheric conditions (15°C), the air exerts a pressure of 14.7 psi (pounds per square inch). This totals out to about 20 tons for the average man, but we don't feel it, because our bodies exert an inner pressure to equalize the situation.

The air is most dense at the earth's surface, because all the air above is pressing down upon it. As we ascend, the pressure lightens rather quickly. At 10,000 ft, it's down to 10.1 psi, which most of us

PRESSURIZATION

can tolerate. But at 20,000 ft, the pressure is reduced to a skimpy 6.75 psi, which is definitely a hostile environment.

Basically, there are two problems to contend with as a plane climbs. One is lack of sufficient oxygen, which, as discussed earlier, will cause hypoxia when we get high enough. The other problem comes from the reaction of the gases in our bodies as the surrounding air pressure decreases.

Several things take place under those conditions. Graham's Law tells us that a gas of high pressure exerts a force towards a region of lower pressure. And according to Boyle's Law, if the pressure on a gas decreases, its volume increases. So as we go up, the gases in our bodies are expanding and trying to get out. If the ascent is rapid, this can cause discomfort in the stomach and middle ear and can possibly bring on a headache as well.

In addition, a considerable amount of nitrogen is dissolved in the blood and other body tissues. With the rapid decrease in outside pressure, the nitrogen comes out of solution, forming painful gas bubbles known as the "bends" when they occur in the joints, and "chokes" when they take place in the chest. These latter problems will not normally occur from an aircraft ascension unless you have been scuba diving shortly before the flight. The rule of thumb is, don't fly to cabin pressure altitudes of 8000 ft or more within 24 hr of scuba diving.

The most common physiological problem occurs in the ears on descent. As the outside pressure becomes higher than that of the middle ear, a partial vacuum is created there, causing the eardrum to bulge inward. This condition is harder to relieve than the reverse situation, since air must be forced up the Eustachian tube to relieve the pressure. The accepted remedy is to pinch your nostrils shut, close your mouth, and blow gently.

The purpose of aircraft pressurization is to supply enough breathing oxygen at altitude and to keep the rate and amount of atmospheric pressure changes within tolerable limits.

How the Plane Is Pressurized. All pressurized planes are powered by either turbocharged piston engines or turbine engines. (It wouldn't make much sense to pressurize a plane if it couldn't get up to altitude.) In a pressurized plane, the turbocharger does double duty. Since it can supply more compressed air than the engine uses, the excess air is routed into the cabin. The so-called "bleed" air goes through a flow limiter called a sonic venturi. Compressing the air causes its temperature to rise greatly, so the air is passed through a heat exchanger, which cools it to within 10 degrees of ambient. Some pressurized aircraft have an adjustable heat-exchanger system that is used to warm the cabin, eliminating the need for a combustion heater.

Once we've brought pressurized air into the cabin, we must let it out again, and it must be very precisely regulated to meet the requirements of both the plane and its human contents. That chore is handled by an outflow valve, which is backed up by an almost identical safety valve.

Cabin Differential. You've probably heard people speak of cabin differential pressure somewhat like this: "This plane has a max differential of 3.35," or "We can maintain a differential of 6.1 psi." What does it all mean?

As the plane ascends, the pressurization system is pumping air into the cabin to maintain as close to sea-level pressure as is feasible. Meanwhile, the pressure outside is decreasing, resulting in a differential in pressure between the atmosphere outside and the atmosphere inside, expressed in pounds per square inch.

The amount of pressure differential an airplane can tolerate is determined by the strength and airtightness of its particular structure. If you took a plane up to 20,000 ft, where the atmospheric pressure is 6.75 psi, and pumped 14.70 psi into the cabin to maintain sea-level pressure, you'd have a differential pressure of 7.95. But light aircraft are not stressed to withstand this differential.

The P-Centurion is certified to a maximum differential of 3.35 psi, meaning you may not pressurize it beyond that amount; that is to say, you may not add more than 3.35 psi to the ambient atmospheric pressure. Let's see how this translates into practical terms.

At the P-Centurion's maximum certified ceiling of 23,000 ft, the atmospheric pressure is 5.94 psi. Add to that the plane's maximum differential of 3.35 psi, and you have a total pressure of 9.29 psi that can be maintained in the cabin. That amount of pressure, 9.29 psi, is found at 12,127 ft, so when the *airplane* altitude of the P-Centurion is 23,000 ft, the *cabin* altitude will be 12,127 ft—which happens to be just short of the FAA's mandatory altitude for supplementary oxygen.

By comparison, Piper's new Malibu was built to have a maximum differential of 5.5 psi. At its maximum operating altitude of 25,000 ft, therefore, the Malibu will have a cabin altitude of only 8000 ft—a significant edge over the P-Centurion.

How Do We Control the Pressure? Keeping ourselves and our passengers safe and comfortable in a flying pressure capsule can be tricky. It requires a combination of sophisticated hardware and good pilot technique; you have to fly the pressurization system as well as the airplane!

PRESSURIZATION   81

The P-Centurion has a simple system, from an operational point of view. (The P-Skymaster has the same type of system, adapted to work off of both engines.) What the pilot sees are three gauges and three controls. The gauges show cabin altitude, cabin vertical speed, and differential pressure. The controls include an on-off switch, a controller knob, and a pressure dump handle. (Fig. 9.2.)

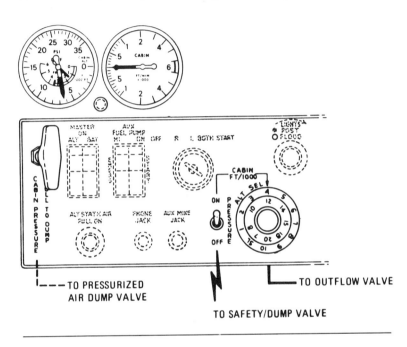

9.2. The Centurion's pressurization system controls and instruments. (*Cessna Aircraft Co.*)

More sophisticated systems include a rate controller, which allows the pilot to adjust the cabin rate of climb or descent by adjusting an additional knob. In the case of the P-Centurion, this is controlled automatically.

The controller knob allows the pilot to dial in the cabin altitude he wants to maintain, and the system will comply—within the limitations of its operating envelope as set by the 3.35 psi differential.

Looking at that knob with its two circular scales in Fig. 9.2, we see that the pilot has dialed in a cabin altitude of 4000 ft, as shown on the outer scale. Notice that the inner scale reads 12. This means the system will maintain a cabin altitude of 4000 ft when the airplane is flying anywhere between 4000 and 12,000 ft; that's the "envelope." Below 4000 ft, the plane will be unpressurized; above 12,000 ft, the system would not be able to maintain a 4000-ft cabin without exceeding the 3.35 psi differential.

To better see how this works, let's look at a typical flight profile (Fig. 9.3).

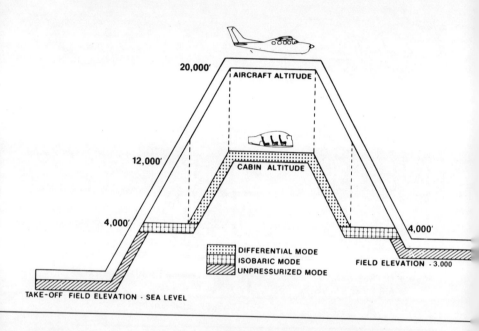

9.3. A typical flight profile. (*Cessna Aircraft Co.*)

You are flying a P-Centurion from a sea-level field to a distant airport with a 3000-ft elevation, and your flight plan calls for a cruising altitude of 20,000 ft. The normal procedure is to dial in a cabin altitude that's 500 to 1000 ft above either your departure or arrival airport, whichever is higher. So you dial in 4000 ft.

As you reach that altitude, the outflow valve begins to close to maintain cabin pressure at 4000 ft, or 12.71 psi. The valve continues to close as you climb, holding that 4000-ft cabin altitude. Of course, the differential pressure is increasing all the while, and when you reach 12,000 ft, which is 9.34 psi, the differential will be just about at the Centurion's maximum allowable of 3.35 psi.

Therefore, as you continue climbing out of 12,000 ft, the outflow valve, which up to now has been obeying the commands of the controller, takes over and starts to open sufficiently to maintain the 3.35 psi differential. This means, of course, that the cabin begins climbing at approximately the same rate as the airplane until you level off at 20,000 ft, at which point the cabin remains at 10,000 ft.

As you approach your destination and begin your descent, the process is reversed, and upon entering the pattern at 4000 ft, the plane becomes unpressurized.

The process is simple, but there are procedures you must follow for safety's sake. First, you need to monitor those three gauges to make sure the system is functioning properly. Second, if you must adjust the controller knob for any reason, do it slowly; you have the power to make the cabin altitude change at an ear-popping rate. Third, remember that the pressurization is being provided by the engine-driven turbocharger, and when you reduce power below approximately 20 in. MP, there goes your pressurization. (This gives the pressurized twin an extra redundancy advantage, since the failure of one engine in a twin will not cause a serious loss of pressurization.) Fourth, remember that the controller thinks strictly in terms of pressure altitude, whereas when you're flying below 18,000 ft, you are using indicated altitude adjusted to the local altimeter setting. Bear in mind an inch of difference between your altimeter setting and the standard pressure of 29.92 in. represents a disagreement of 1000 ft between your altimeter and the controller's brain, so you may have to compensate to keep from landing pressurized.

As I indicated earlier, virtually all pressurized planes other than the P-Centurion and P-Skymaster have a rate controller, an additional knob that allows the pilot to select the rate, in feet per minute, at which the cabin will climb or descend. This gives you something else to twiddle, but it's a small amount of workload that enables you to smooth out the system.

You'll see this clearly if you take another look at Fig. 9.3. Using a rate controller, you can bring the cabin up or down at a *continuous* rate rather than the *steps* required by the P-Centurion's more elementary system. The difference is one of comfort; you may feel some slight pressure bumps when the latter system shifts modes.

What's Coming? Garrett now makes a computerized controller, which is being installed on some of the business jets. Presumably, a version of it will become available to the piston market in the near future. The pilot simply dials in the barometric pressure at the airport of departure, plus the field elevation at the destination, and then, while enroute, he punches in the destination's barometric pressure.

The computer sets the cabin climb and descent rates automatically for optimum comfort.

That's a nice piece of sophistication. But the big news is the new single-engine pressurized aircraft that are coming to market. Granted, they're six-place airplanes and they're expensive. But we have some very capable four-place turbocharged planes in the general aviation fleet — and who knows, perhaps some day *they'll* be pressurized, too.

10. Anti-icing and De-icing Equipment

How to prevent ice...How to get rid of it...Which planes have "known-icing" systems...Alternatives to boots.

AS PLANES become more capable, and as they are utilized more for business as well as pleasure, they are exposed to a greater variety of weather conditions—including icing. As a result, anti-icing and de-icing equipment is being made available for an increasing number of aircraft models, both as factory installations and as retrofits.

In this chapter, we'll consider the various types of equipment, the aircraft for which they are available, what the equipment will and will not do, and tips on operation and maintenance. We'll also look ahead to future developments.

Anti-ice and De-ice. Most fully equipped planes have both anti-icing and de-icing protection. Anti-icing equipment is designed to prevent ice from forming, and that's usually accomplished by heating the surface where icing conditions are expected to be encountered. Typical anti-icing equipment consists of electrically heated props, windshield, pitot tube, and, in the case of turbine-powered planes, engine inlets. Obviously, this equipment can also serve a de-icing function if ice has already formed, but it is most effective when used before ice accumulates.

The function of de-icing equipment is to remove ice after it has formed. Pneumatic wing and tail de-icers fall into this category. They consist of boots with inflatable tubes installed on the leading edges of the wings, horizontal stabilizer, and sometimes the vertical stabilizer. (Fig. 10.1.)

Ice protection equipment for a single-engine plane can run from about $4000 for a windshield and prop system, to about $20,000 for a complete airframe package certified for flight into known icing.

Just What *Is* "Known Icing"? If you receive a weather briefing that includes a pilot report of ice encountered at a certain altitude, that report constitutes "known icing," and if you fly at that altitude in

10.1. Considering their replacement cost of about $4000, it pays to protect your boots from the ravages of the sun. (*Keith Connes*)

an airplane that is not certified for flight into known icing, you are in violation of the FARs.

So much for the legalities. Now, let's take a look at the various categories of icing severity, as defined in the *Airman's Information Manual:*

Trace: Ice becomes perceptible. The rate of accumulation is slightly greater than the rate of sublimation. It is not hazardous, even though de-icing and anti-icing equipment is not utilized.

Light: The rate of accumulation may create a problem if flight is prolonged in this environment over 1 hr. Occasional use of de-icing and anti-icing equipment removes and prevents accumulation.

Moderate: The rate of accumulation is such that even short encounters become potentially hazardous, and use of de-icing or anti-icing equipment and flight diversion is necessary.

Severe: The rate of accumulation is such that de-icing and anti-icing equipment fail to reduce or control the hazard. Immediate flight diversion is necessary.

Bear in mind that, just as with turbulence PIREPS (pilot reports), one man's "moderate" may be another man's "severe" — depending on the equipment being flown, the pilot's capability, and subjective judgment. So be conservative when evaluating reports of icing conditions.

What the "Big Four" Offer

CESSNA. The smallest Cessna on which you can get any kind of factory-installed airframe ice protection is the Turbo Skylane. This plane, along with the Turbo Skylane RG and the Stationair Series aircraft, can be equipped with prop and windshield anti-ice systems.

Complete known-icing packages are available for the Turbo- and Pressurized-Centurions and all twins.

PIPER. Piper has a full ice protection package available for the Turbo Saratoga SP and Malibu, as well as all twins.

BEECH. The only Beech single to offer any icing equipment is the Bonanza, which comes with a hot-prop option for either the two-blade or three-blade propeller; a 100-amp alternator is required for this installation. According to a sales representative, Beech takes the position that flight into icing should not be made without the redundancy features of a twin, and the hot prop is offered "to get the pilot out of trouble rather than get him into it."

The Duchess has no factory-installed ice-protection equipment. However, this model has been retrofitted in England with a TKS porous-panel system, which is described later in this chapter.

The various Barons and the Duke have full known-icing packages as options; the equipment is standard on the King Airs, which additionally have a heated-brake option to prevent slush from freezing and locking the brakes while the plane is parked.

MOONEY. Both the 201 and 231 offer a hot-prop option. Mooney does not provide optional leading-edge boots, citing the weight penalty as the reason. The pressurized 301 will have full ice protection, but it may not be the conventional equipment currently in use. Mooney is examining alternate developments, some of which are still in the experimental stage.

How to Use the Equipment. If there's a possibility of encountering ice during your flight, make sure your ice-fighting equipment is in good operating condition—while you're still on the ground. The name of that tune is "preflight," and unfortunately anti-ice/de-ice equipment isn't on all the checklists it should be. Turn on your heated pitot and stall warner vane, and feel them during your walk-around. Check the de-ice boots for cracking. Then, during the run up, turn on the windshield heat and feel that. Note: on some aircraft, it is not advisable to operate the windshield heat with the engine off; check your information manual carefully for the correct procedure. Cycle the de-ice boots and check them visually, along with the vacuum gauge. Cycle the prop heat and check the ammeter.

The technique of using the de-ice boots is as follows: after about $1/4$ in. of ice has accumulated on the leading edges, the pilot presses an electrical switch, causing the rubber tubes to inflate and deflate by means of an engine-driven vacuum pump—or, in the case of a turbine-powered plane, a turbine compressor.

This expansion and contraction of the rubber tubes is designed to break up the ice, which is then carried off in the slipstream. Some jets use an anti-ice system, warming the leading edges by bleed heat from the engines.

Now that you know all your fancy gear is putting out, are you ready to bore through any and every kind of icing condition? The answer is a resounding "No!" It should be emphasized that *there is no ice-protection equipment made that will work under all icing conditions.* Every professional pilot I've discussed this with agrees that the purpose of ice-protection equipment is to buy time so you can get *out* of icing conditions.

The first ice-avoidance procedure comes during your weather briefing. Ask for the forecast freezing level along your route of flight, plus any PIREPS on icing. If you will be passing through a frontal system, make a note whether it's a warm front or a cold front.

Turn on your anti-icing equipment any time you are in visible moisture and the temperature is +4°C (+39°F) or colder. If ice begins to form on the plane, note whether it is clear or rime. Clear ice is slick; rime ice is rough and white. If it's clear ice, your best move is probably a 180. Regardless of the type of ice, if you're continuing in it, don't sit there and wait to see what it will do. It will probably get worse, by which time your options will have deteriorated along with your plane's performance. Immediately ask for another altitude. Under most conditions, you'll probably want to go up, because if that doesn't work, you can always go down — whereas if you go down first and keep picking up ice, you may not then be able to go up.

The exception to the go-up-first rule of thumb can occur when you are in a warm front; your angle of climb may equal the slope of the front, which would have the effect of keeping you in ice. Know your weather conditions, and monitor the outside air temperature as you change altitude. Write down the temperatures you encounter at various altitudes, to give you an idea of what to expect higher or lower.

Do not cycle your de-icing boots until you've got an ice build up of ¼ to ½ in., because they will not do the job until there is that amount of accumulation. Furthermore, if ice starts to form on the boot while it's in the inflated stage, you may not be able to get it off. So this is one instance where you don't hit the panic button at the first sign of ice.

When you ask for a change in altitude, be sure to mention the word "ice." This should get you quick service. However, if there is an undue delay, and the ice keeps building, you'll probably want to exercise your pilot's authority and explain to ATC what you are going to do. Most icing conditions are within a maximum range of 6000 ft,

and a change of 3000 to 4000 ft will usually get you out of it.

Try to avoid flying through the tops of cumulus clouds, where icing is often at its worst. If you must penetrate a front, determine its position and adjust your route of flight if necessary to get through it in the shortest period of time, rather than running parallel to it. And of course, keep abreast of conditions at your destination; if there's the likelihood of picking up additional ice during your approach, start checking your alternates.

Maintaining Your Equipment. If you think $20,000 for a complete ice-protection package is high, consider this: the average life of today's neoprene boots in noncommercial service is about six years, *if* they are properly maintained. Since the cost of replacing those boots runs upwards of $5000 and requires four to five days downtime, it certainly pays to provide that maintenance.

B. F. Goodrich, which makes most of the anti-ice/de-ice equipment in use today, advises that the greatest enemies of rubber or neoprene are ozone and the ultraviolet rays of the sun. There are a number of products intended to protect rubber from these elements, but the only one Goodrich recommends is Age-Master, available through their distributors or possibly your local variety store. Goodrich does not make Age-Master, but they have tested it along with other products. They have found that many others have a water base, will wash off easily, and, they say, may do more harm than good. Age-Master, on the other hand, has a petroleum base and is absorbed into the boot material.

To make the boots slick and thus help prevent ice formation, Goodrich makes a silicone-based product called Icex, which should be applied every 150 hr. Finally, if you want to add more shine and a little more protection, you can apply paste wax, but this is mostly a cosmetic benefit.

Future Developments. A new method of boot installation is being utilized on the Lear Fan. The airframe's composite construction allows the boots to be built into the leading edge as a removable panel. At replacement time, the panel can be replaced by a new unit at a considerable saving in time and labor. (A good part of the labor cost of replacing conventional boots goes into the time-consuming process of peeling off the old boots, which are cemented to the leading edges.)

A fairly recent improvement in prop boots is Goodrich's Hot-Prop, which uses an etched heating element instead of conventional wires for greater efficiency and durability. HotProp also has a two-cycle system that is simpler than previous four-cycle systems.

As noted earlier, there's an alternative type of anti-ice/de-ice equipment that has been in use in Great Britain for some time and is now being certified for some aircraft in this country. It's a porous-panel system made by a company called TKS Ltd. (Fig. 10.2.)

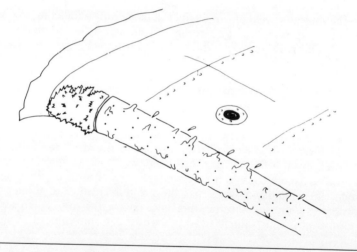

10.2. The TKS porous-panel system.

Instead of boots, the system uses mesh-screen panels attached to the leading-edge surfaces. In icing conditions, a glycol formula is pumped through the mesh, functioning either as an anti-ice system to prevent ice formation or as a de-icer to dispose of accumulated ice. The glycol is also used on the props and windshield.

TKS claims that the system has the following advantages over de-icing boots: less effect on the aerodynamics of the wing airfoil; a greater area of protection, since the fluid can run back over the top and bottom surfaces; simpler operation and, therefore, reduced chance of pilot error.

An obvious drawback is that the system's effectiveness is limited to the amount of glycol carried on board, which in turn becomes a weight-limiting factor. For example, consumption on one light twin is 13 lb/hr for the leading edges and 11 lb/hr for the propellers. Operators have to make a judgment about how much ice protection they want to carry as a trade-off against payload.

As of this writing, the TKS system has been approved for the Citation S/II, and the U.S. distributor is working on certification for the Skylane, Centurion, 310, Bonanza, Duchess, and Baron. Initial cost will be about 20 percent higher than boots, but the TKS system could be more economical in the long run because of the replacement cost of the boots.

Other methods of ice control are conceivable. For example, it is possible someone will design a method of "packaging" and routing heat from reciprocating engines, as is now done on some jets.

And how about this: at least one research lab is trying to use supersonic waves to cause water droplets to freeze *before* contacting the airplane's skin—which sounds as if it could be the ultimate anti-ice system.

But for now, no matter how your plane is equipped, the old-fashioned advice still holds true:

1. Stay out of ice.
2. And if you do get into it, get out of it.

11. Avionics

The Big Three and their digitals...Navcomms, RNAV, Radar, Stormscope, Loran, etc....Factory or field installation?...How to keep your avionics healthy.

(Texas Instruments)

IT'S NO SECRET the most innovative products that have come out of general aviation in the past decade have not been airplanes, but avionics.

The majority of the single-engine production planes and their engines can trace their lineage back to the 1940s (the Bonanza made its first flight on December 22, 1945), 1950s, and 1960s. The only real excitement since then has come from the designers of homebuilts, sailplanes, and ultralights.

The avionics industry has taken advantage of the large-scale integration and microprocessor technology developed primarily for much larger markets. The benefits to us, the end users, include:

1. Big capability in small boxes. King's KNS 80 is a good example. In a single 6-lb box, 6.31 in. wide × 3 in. high, is a 200-channel VOR/localizer receiver, a 40-channel glideslope receiver, a 200-channel DME and a 4-waypoint digital RNAV receiver.

2. Lower parts count for a given capability. This pays off in less heat, greater reliability, and easier serviceability.

3. Memory. An obvious benefit is the ability of a navigation or communications radio to store frequencies, or an RNAV or Loran receiver to store waypoints, greatly reducing pilot workload. For exam-

AVIONICS

ple, ARNAV's R-21/NMS Loran receiver stores the latitude and longitude of nearly all airports and VORs in memory—that's more than 9000—and has room for another 150 user-designated waypoints.

The visual hallmark of the now-generation radios is the electronic digital display, where mechanical numbers and analog needles have been replaced by agile bars of light that can form rapidly-changing numerals. There are four different types of digital displays in general use:

1. Gas discharge is probably the most widely used in today's light-plane radios. It requires a high voltage, but is quite reliable. King had some leakage problems early on, but the manufacturer says they have been corrected.
2. Light-emitting diode is probably the most reliable display, but some formats have poor visibility in direct sunlight. The high intensity dot-matrix LED displays overcome this handicap.
3. Incandescent has the best visibility but generates heat and is expensive to produce.
4. Liquid-crystal display has had problems in the past. It degenerated at low temperatures, would reflect light-colored objects such as white shirts, and was hard to read when viewed at an angle. However, the temperature and reflective problems have been overcome, and the viewing angle can often be adjusted to permit easy readability.

The Manufacturers. We can't hope to cover every avionics manufacturer, let alone every product, but we'll look at the most significant companies and the black boxes you'll be most likely to use. Let's start with the Big Three manufacturers of panel-mounted radios.

KING. This organization bears the stamp of its imaginative and aggressive founder, Ed King, whose first company was bought by Collins Radio in 1955. King stayed on as president and tried to talk Collins into building products for the general aviation market. Collins, which was then focusing on the military and the airlines, was not interested, so King left and formed King Radio Corporation.

At the time, the leader in the general aviation field was Narco, which had a strong foothold with the aircraft manufacturers. Like any newcomer, King had to prove himself by appealing to the aftermarket. He accomplished this by producing the first crystal-controlled 1 + ½ system (a navigation and communications radio combined in a single box).

Narco faltered during the 1970 recession, while King forged ahead. Today, King Radio has a very broad range of products and is the unquestioned leader in the light-plane avionics business.

King makes a Gold Crown line of remote-mounted radios for corporate aircraft and a Silver Crown line of panel-mounted avionics for smaller planes. The Silver Crown line includes navs, comms, nav/comm systems, DMEs, RNAVs, ADFs, transponders, autopilots, encoding altimeters, radar altimeters, weather radar, a radio telephone, and a variety of indicators and audio panels.

Early in 1985, King Radio Corporation was acquired by Allied Corporation's subsidiary, Bendix Aerospace. The new company name is King/Bendix.

NARCO. As mentioned earlier, Narco was the leader in light-plane avionics prior to the recession of 1970. Considering the nightmarish decade that followed, it's a wonder Narco is still around. King kept topping them with imaginative new products. Narco did produce the very successful Mark 12 transceiver, but replaced it prematurely with a series of models that didn't work very well. Cessna ceased to be a customer when they decided to offer only their own ARC brand of radios as factory equipment. (An exception was Narco's DME 190/195, which Cessna continued to buy because ARC did not make an equivalent unit.) Radio shops were unhappy with Narco quality and the fact that the brand was being widely marketed at discount prices by mail-order houses.

The final blow came when Narco's parent company, whose main interest was in the medical field, put its avionics stepchild on the block—where it lingered in humiliating uncertainty for a year. Rescue came at last in the form of Ed Zimmer, a high-tech industrialist and aviation enthusiast who likes to buy ailing companies.

At about that time, Narco was further floundering in an ill-conceived effort to sell a high-priced line to the business aircraft market. Zimmer made some management changes, and Narco went back to the single-engine/light-twin market it knew best, with the beginnings of an all-new Centerline II model line.

Recognizing that they had to overcome some skepticism, Narco offered a 60-day no-questions-asked replacement guarantee, in addition to the standard one-year warranty, plus an optional extended warranty. The Centerline II equipment was well received. Narco has won its way back on the options list at Piper and Mooney, and Narco products can be special-ordered on Beech aircraft as well.

COLLINS. The original Collins Radio Company was founded more than 50 years ago and during its early years concentrated on equipment for commercial broadcasting, police radio, and amateur rigs. The company became part of Rockwell International in 1973.

Their avionics production has always had its emphasis in the military, airline, and corporate markets. The remote-mounted radios are called Pro Line, and the panel-mounted avionics are named Micro

Line. Although there have been some innovative Micro Line developments spaced rather widely over the years, Collins does not appear to be interested in joining the race by King and Narco to turn out a spate of digital products for the single-engine and light-twin markets.

Ironically, one of Collins's most enthusiastic attempts to cater to the lower-end market ended as an embarrassing failure. In 1979, with much fanfare, they introduced a device called the DCE-400. The initials stood for distance computing equipment, and it was apparently intended as an ersatz DME.

The DCE-400 computed radial information from the plane's two nav receivers to provide readouts on distance, groundspeed, and time-to-station. It worked well enough, but had the inherent limitation of being dependent on usable signals from two offset VORs, which was often no problem at altitude, but could be a definite problem when you got down low—such as on approach, when you most wanted the information.

The distance computing equipment was neither fish nor fowl. It did what a DME does, but not all the time, and it required much more pilot input than the real thing. At its price of about $1400, the DCE-400 was really not so much a poor man's DME as a rich man's VFR gadget. When last heard from, it was being remaindered for several hundred dollars by surplus dealers.

Now that we've met the manufacturers, let's meet the products.

Navs and Comms. You can get VHF (very high frequency) navigation and communication equipment either combined in a single nav/comm box, sometimes referred to as 1 + ½ system, or in separate nav and comm sets, called a 1 + 1 system. The former takes up less panel space and costs less, while the latter offers more redundancy. Some nav sets are sold with built-in glideslope receivers and some without. Virtually all modern comm units have 720-channel capability; the nav units have 200-channel capability; and the glideslope receivers have 40-channel capability. Prices for the combination nav/comm units start at about $3300, including course deviation indicator, while the separate nav and comm and indicator systems are close to $5000. A low-budget alternative is the shared receiver system, a box that lets you navigate or communicate, but not both at the same time. This type of radio, with built-in indicator but no glideslope receiver, is available for as little as $1600.

King's nav/comm systems are the KX 155 and KX 165. Their digital navs and comms have the flip-flop frequency-management feature, with an active window and a standby window. The desired frequency is entered in the standby window. When the pilot wants to activate it, he presses a button, and the active and standby frequencies trade places. If the pilot wants to go back to the previous frequency, he presses the button, and it flip-flops back again. The two models are similar, but the KX 165 has a "radial" feature not found on the lower-priced KX 155—pull a knob and a digital display of the VORTAC radial you're on appears in the standby frequency window. (You cannot choose "To" or "From"; it's "From" only.) Transmitter output on either model is 10W nominal.

The KY 196/197 is King's flatpack comm transceiver. Like other King products, it uses the EAROM (electrically alterable read only memory), a circuit component that stores the active and standby frequencies in memory through shutdown, so they'll reappear when the power is turned on again. The EAROM chip does not use the aircraft's electrical system or standby batteries. The KY 196 transmits a minimum of 16W at 28V and the model KY 197 puts out 10W minimum at 14V.

This transceiver has the flip-flop arrangement with one frequency in the active mode and one frequency in storage. In the latest versions, designated KY 196-05 and KY 197-05, King has added a fillip to the flip-flop: a memory bank that will store nine additional frequencies. That could be useful if you fly out of an airport that has ATIS (automatic terminal information service), clearance delivery, ground control, and tower. You can program those frequencies in and add those you expect to use enroute, plus the ones that are waiting at your destination. Then all you have to do is figure out a system that will help you remember which frequency you've programmed into which channel.

There's also a KY 196-10 and a KY 197-10. These versions have a provision for a remote switch so you can whip through stored frequencies without your hand ever leaving the yoke (or cyclic, if your wings go 'round). Incidentally, if you have an older KY 196 or KY 197, the local King dealer can modify it to incorporate the additional frequency storage of the new units.

Anyone buying a radio with a digital display has to think about what happens if the lights go out. Many sets are built with some provision that will let you tune "blind" if necessary. In the case of the KY 196, there is a direct-tune mode. If you turn the volume control to the "Off" position, hold the frequency transfer button in, and then turn the volume control back on, the radio will go to 120.00 MHz. So if you've lost the display, you can count the clicks as you turn the frequency selector knobs, and that should keep you in business. In this mode, you will be tuning directly into the "Use" window.

The KN 53 is the KY 196/197's companion flatpack nav receiver.

Narco's low-end nav/comm, the Escort II, has a shared receiver, like the old Escort I, but there the resemblance ends. First of all, the entire radio, including indicator, fits into a 3-in. panel hole. Furthermore, the Escort II has a gas-discharge digital display for the nav and comm frequencies and for the OBS (omnibearing selector) readout. In addition, it has an ECDI (electronic course deviation indicator) rather than the conventional mechnical needle.

The Escort II has a digital RMI (radio magnetic indicator) feature that works as follows: the OBS knob has three positions. When the knob is in the center position, the CDI (course deviation indicator) functions normally. If you pull the knob out, the left/right digital meter will automatically center and give you a heading to the VOR station to which you're tuned. If you push the knob all the way in, the left/right meter will disappear, and you'll get a digital readout of your bearing from the station.

The Escort II will receive VOR and localizer signals, but it does not have a glideslope receiver. The transmitter has a power output of 5W nominal.

Narco's 1 + ½ system, the Mark 12D, features flip-flop frequency management, a *T* that lights up when you're transmitting, automatic gain leveling, and automatic squelch (Fig. 11.1). A "keep-alive" memory option is wired to the aircraft battery; this retains in memory the last frequencies in use even when the power is shut off. Transmitter power output is 8W nominal.

11.1. The Narco Mark 12 D nav/comm. (*Narco Avionics*)

The COM 810/811 1 + 1 system has flip-flop with a difference. Where the Mark 12D allows you to enter frequencies only on the standby side, the COM 810/811 has a selector switch that permits you to enter frequencies on either the active or standby side. An arrow on the display points to the side armed for entry. Narco feels the ability to change frequency on either side offers greater flexibility and might be more comfortable for the pilot who has been raised on mechanically tuned radios and is used to direct dialing. The downside is that it offers the pilot another opportunity to goof up by having the switch in the wrong place at the wrong time. But if it becomes a problem, the user can always tape the switch to the position that suits him best.

The unit has the hot-wired "keep-alive" memory. If you prefer not to have the constant 1.0 mA draw on your battery, another option will enter 121.5 on the active side and 121.9 on the standby side every time the power comes on. These frequencies will also be entered automatically if the display should fail, and you can then change frequencies by counting detent clicks.

Other features include variable-rate tuning which helps you make large frequency changes quickly, and automatic audio leveling.

The 810 is 14V and the 811 is 28V. Transmitter output is 8W nominal.

The companion set is the NAV 824/825. Unlike the COM 810/811, its flip-flop feature allows frequency changes only on the standby side. Narco's rationale is if you inadvertently change from one VOR to another on the active side while the autopilot is coupled, the plane might try to do a split-S or something.

The COM 824/825 has a digital bearing readout, which can be called up in place of the standby frequency. You can select "To" or "From." If you want to enter a frequency while in a heading mode, just start dialing, and the display will revert to standby and stay there for 4 sec after you've stopped turning the knobs; then it will return to the bearing display.

The difference between the NAV 824 and the NAV 825 is that the former does not have a glideslope receiver; the latter does.

It should be noted that Narco has now gone completely over to the rectilinear-course deviation indicators. That is, the entire needle moves sideways or up and down, rather than the "windshield wiper" movement of past CDIs.

Collins led the way in panel-mounted digital navs and comms with memory. Their Micro Line VHF-251 comm transceiver allows you to store and recall a standby frequency. However, unlike the flip-flop designs, it displays only the active frequency. Transmitter output is 10W nominal.

The companion VIR-351 nav receiver offers a digital readout on either the "To" or "From" bearing of the VOR or waypoint in use. The two radios are produced in a rectangular format, somewhat like the panel units of remote-mounted equipment. They are available with faceplates colored green, red, blue, brown, beige, or black.

Collins's later-model VHF-253 comm set was designed specifically for the helicopter market, but its features should have considerable appeal for fixed-wing pilots. For example, radios don't thrive too well on the vibration prevalent in helicopters. So the 253 has been given a die-cast case that looks like a Mosler safe compared to the packaging of most general aviation radios. That should help increase the 253's lifespan in fixed-wing aircraft as well.

The VHF-253 has flip-flop frequency management, with storage capability of six frequencies. The display is LCD, but instead of the usual grey background, Collins uses gold for better visibility. The main advantage of LCD is that it requires very low voltage; in fact, Collins says that the 253's display draws one-millionth the power of the commonly used gas-discharge readouts. The payoff comes in a substantial reduction in heat, which is another step forward in reliability.

In the past, a major problem with LCD has been its tendency to congeal at low temperatures. Collins experimented for several years and says the 253 display will operate continuously at temperatures ranging from $-20\,°C$ to $+55\,°C$ ($-4\,°F$ to $+131\,°F$). Another challenge that had to be overcome was reducing reflection, sometimes known as the "white shirt" effect.

Transmitter output is 10W nominal. And the companion nav unit...does not exist. Collins was planning to produce one, but the general aviation slowdown caused them to put the project on the back burner. This has hurt sales of the 253, particularly since it is in the flatpack format and doesn't match the VIR 351 at all. Granted, if you stack two 253s and put a brace of 351s side by side, both pairs of radios will occupy space of equal dimensions; but most buyers like to have their navs and comms match each other cosmetically as well as operationally.

Perhaps by the time you read this, there will be a companion nav set to the 253, but at this writing Collins does not seem to be interested in developing additional products for the panel-mounted market.

Terra's TXN 960 nav/comm system has mechanical frequency readouts and a built-in ECDI that uses gas-discharge light bars instead of mechanically driven needles to indicate course deviation. It works like this: if you are tracking a VOR radial and the presentation shows two light bars in the center of the display, you are on course. If you drift, say, to the left of course, light bars will radiate in a horizontal line from the center of the display to the right—the further off course, the more bars. An arrow will point to the right to emphasize the message. There are vertical bars for the models with glideslope capability.

The ECDI is offered in a choice of flavors. One can display two

VOR/LOC course and one external Loran C course deviation indication.

Individual nav and comm units, as well as the ECDI, can be purchased separately. Nav transmitter power is 5W nominal.

Aire-Sciences, formerly Edo-Aire, makes a line of mechanically tuned nav/comms. Their RT-553A is a shared-receiver box with built-in CDI. The RT-563A is also a compact single-box unit, but with separate nav and comm receivers. A feature of this unit is Auto-Omni: when the OBS knob is pushed in, the azimuth dial of the set automatically rotates until the needle is centered on the "To" bearing of the VOR station in use. This saves manually centering the needle. Transmitter output is 6W on both sets.

Aire-Sciences makes separate nav and comm boxes as well. In addition, the company offers mechanical CDIs and an ECDI. The latter has an LED display and includes a digital RMI. Another feature is a "course-two" memory that allows the pilot to store a missed approach course, which is available for instant recall if needed.

Sperry purchased Cessna's ARC line of radios and autopilots in 1983 and moved the factory from Boonton, New Jersey, to its own facilities in Phoenix, Arizona. The equipment continues to be manufactured primarily for factory installation in Cessna aircraft, but it is available also to the aftermarket.

There are three product lines: the 300, 400, and 1000 series. (There is also a 200 autopilot.) The 300-series radios generally go into the light singles, while the 400s are installed in the high performance singles and light twins, and the 1000 series is aimed at the corporate aircraft.

All three series offer digital nav/comms; however, the 300 nav/comms do not have memory. Their CDIs have automatic radial centering as an option. This is similar to the Auto-Omni feature of Aire-Sciences, except it can automatically center the needle "From" as well as "To."

Radio Systems Technology offers a choice of assemble-it-yourself nav/comm kits. The RST-571 has 360-channel comm and 200-channel nav capability, with thumbwheel tuning. The RST-572 provides 720-comm channels. Both models are priced well under $1000. They have a rather primitive looking built-in CDI; two lights serve as "To" and "From" indicators, which is a good approach, in that it calls attention to station passage. Future models are expected to drive a glideslope indicator.

The kit manufacturer states that the entire radio may be constructed in as little as 40 hrs. When the set is finished, the customer can then send it to RST for a free initial alignment and certification.

DME. Distance measuring equipment is a pulse system that interrogates a selected VORTAC and translates into distance the difference between the time it takes to reach the VORTAC and the time needed to receive a reply from that station's transponder. Modern DME sets have computers that can show distance to or from the station, groundspeed, and TTS (time-to-station). If the unit also has a built-in clock, it can further translate TTS into ETA (estimated time of arrival).

DME is a requisite for VOR-based RNAV systems. DMEs are priced in the $2000 to $5000 range.

King's KN 62A is a flatpack panel-mounted DME. It can be tuned directly and channeled through most nav receivers, thus providing the pilot with two frequencies at once. The display can be selected to read either distance and frequency or distance, groundspeed, and time-to-station. The unit has 100W nominal output.

The KN 63 is a remote-mounted DME, which is channeled selectively from Nav 1 or Nav 2. It also has a "Hold" mode, where it will remain channeled to the last selected frequency even when you are changing nav frequencies. The indicator shows distance, speed, and time-to-station. Nominal ouput is 100W.

Narco's DME 890 can be tuned directly or channeled through a nav receiver, allowing two frequencies to be available at all times. The display can be selected to show either distance and frequency or distance, groundspeed, and time-to-station. When the unit locks on a station, the groundspeed indicator starts counting from 120 kn rather than zero, for a faster reading. If it goes off line, the DME's memory will maintain the display for 10 sec.

Narco also makes a unit known as the IDME-891, which provides DME capability at about half the cost of other low-priced DMEs. How is this possible? The key word is *integration*. The manufacturer has combined into one black box a DME and a VOR/LOC/GS/marker beacon indicator — and it all fits into a standard 3-in. instrument hole. This saves panel space as well as money. There are two provisos necessary to take advantage of the IDME's benefits. One, of course, is you must have a need for a new glideslope indicator as well as a DME, since you can't get one without the other. Two, you must have one of Narco's new-generation nav receivers, such as the Mark 12D, the 824/825, or the RNAV 860, because the IDME is not compatible with any other make of nav receiver — or even with earlier vintage Narco radios.

Collins's DME 451 displays ETA, elapsed time, and Greenwich Mean Time, in addition to the normal DME functions. It has a groundspeed readout of up to 400 kn and time-to-station of up to 120 min.

ADF. The automatic direction finder points at stations that broadcast in the low-frequency range, such as NDBs (nondirectional beacons) and AM radio stations. Compared to practically any other type of radio navigation, precise ADF tracking is demanding and requires periodic practice. However, when low ceilings force you below line-of-sight VOR reception, it's comforting to have ADF equipment on board. Also, the IFR pilot needs it for those airports that have only an NDB approach.

ADFs are priced in the $2000 to $3000 range.

King calls its KR 86 a digital ADF system, which may confuse those who equate "digital" with "electronic." The KR 86 has a mechanical tuning system, but it is crystal controlled and clicks in the frequencies precisely, rather than sweeping the band as the earlier sets did. The KR 86 is an all-in-one unit, with the indicator built into the box.

The KR 87 is of the electronic digital generation, with flip-flop active and standby frequency selection and a remote indicator. Also included are two timer functions. One is a flight timer that starts automatically when the set is turned on. The timer can be rigged to a squat-switch or other device so it will start counting when the plane is airborne. There is also an elapsed timer that provides a stopwatch function and can be set to count up or down. When the timer displays are in use, they replace the standby frequency, at which time the pilot can bypass the flip-flop feature and direct-dial a new frequency. This enables the pilot to scan the dial for local radio broadcasts.

Narco's ADF 841 has flip-flop frequency management, a flight timer and an elapsed timer, and a remote indicator. It functions in much the same way as the competitive King KR 87, except the KR 87 has pushbutton controls and an EAROM memory circuit, while the ADF 841 uses knobs and a hot-wired "keep-alive" memory circuit.

Transponder. The transponder is probably the least exciting piece of avionics in your panel, but King has brought out a model with some bells and whistles and a price tag of over $2000, which is more than double the price of a plain vanilla set. This model, the KT 79, is the first panel-mounted transponder with all solid-state construction, eliminating the cavity tube that has been one of the transponder's most maintenance-prone components (Fig. 11.2).

The unit has two digital readouts. One is the Mode A code, or squawk, you've dialed in. The other is the altitude shown on your encoding altimeter, assuming you have one on board. The altitude is displayed as FL (flight level) in hundreds of feet, and it is referenced to the encoder's setting of 29.92 in. Hg. An obvious benefit is it serves as a backup in case your regular altimeter fails, provided you correct as necessary for the 29.92 setting.

AVIONICS

11.2. The King KT 79 transponder. (*King Radio Corp.*)

Another feature is the VFR button, which activates a stored frequency. Normally, you will program the 1200 code into memory, so when ATC tells you, "Radar service terminated, squawk VFR," push the button and you've got it.

RNAV. VOR navigation has been greatly enhanced by RNAV (area navigation) equipment. An RNAV system uses a built-in computer that enables you to "displace" the signals of a VORTAC from that station's location to another location of your choice, by dialing in the desired coordinates, i.e., bearing and distance. You can then navigate to the "phantom" VORTAC, or waypoint, as if the station were actually situated at the new location. The CDI or HSI (horizontal situation indicator) needle will give you bearing information, and the "To" flag will change to "From" as you pass over the waypoint. A DME is required for RNAV, and if it's a current model it will give you distance and groundspeed to the waypoint.

Note: both VORTAC and VOR/DME facilities can be used to create waypoints. For simplicity, I am using the term *VORTAC* in reference to both types of facility. Here are the major benefits of RNAV:

1. You can often place your waypoints on a straight line to your destination and avoid zigzagging from VORTAC to VORTAC.

2. You can put a waypoint at the approach end of the runway at your destination airport.

3. When making a precision approach, you can place a waypoint at the outer marker; it's very useful to know your relationship to that important fix when being vectored here and there by ATC.

4. You can create a "fence" by placing a waypoint at some place you don't want to transgress, such as a mountain that is higher than you are. This is an extra measure of safety when you are getting ATC vectors.

On RNAV sets, your course deviation when navigating to a waypoint is displayed on the CDI or HSI as *linear* rather than *angular*. For example, in the RNAV/ENR (enroute) mode of King's KNS 80, each dot on the CDI or HSI represents one nautical mile of course deviation, rather than the two *degrees* expressed in the VOR mode. This gives you more useful information, since if you're flying to one side or the other of a VOR radial, what you really want to know is how far off track you are in *miles*.

You can track a VORTAC with this same linear method by switching to the "VOR parallel" mode. For example, if you want to maintain a parallel track 5 NM to the left of a VOR airway, use "VOR parallel" and fly a heading that will keep the CDI needle five dots to the right.

For greater sensitivity when making an RNAV approach, the RNAV/APR (approach) mode will give each dot the value of ¼ NM of course deviation.

RNAV units suitable for light aircraft are priced in the $3000 to $7000 range. Some sets include a built-in DME, but the majority do not.

King's popular KNS 80 is not merely an RNAV, but an integrated navigation system that includes a VOR/LOC receiver, a glideslope receiver, and a DME (Fig. 11.3). The RNAV has four-waypoint capability.

11.3. The King KNS 80 navigation system. (*King Radio Corp.*)

Because both DME and RNAV information is shown digitally on one box, there's not room to display all of it at once. Consequently, the DME data—distance, groundspeed, and time-to-station—are displayed simultaneously, but VOR/LOC frequency, waypoint radial, and waypoint distance have to be called up one at a time. Angular or linear deviation can be selected at the push of a button.

With so much capability combined in one box at an attractive package price, the KNS 80 has sold very well in both the original equipment and retrofit markets. However, the all-in-one concept brings with it a couple of possible disadvantages that should be considered. One, the box usually goes into the center radio stack, putting the DME readouts somewhat out of the pilot's direct line of vision. Two, if the unit has to go into the shop, you lose a lot of nav functions in one fell swoop.

Now let's move along to a fancier model, the KNS 81. This is priced about the same as the KNS 80, but the final cost is higher because the box does not include a DME. As indicated above, there is an advantage to this, in that a separate DME, such as King's compact KN 63, can usually be installed closer to the pilot's scan than a larger multipurpose box. Another benefit is the separate DME can be channeled both to the KNS 81 and to a second nav receiver. (The DME built into the KNS 80 cannot be channeled to another nav receiver.)

The KNS 81 has nine waypoints, as compared with the KNS 80's four waypoints. On the KNS 81, you can see all three waypoint parameters—frequency, radial, and distance—displayed at once. On the KNS 80, you have to punch these up one at a time.

The KNS 81 has a "radial" button that will display on the DME indicator the radial you are on separately from the active VORTAC station or waypoint. And if you are navigating on a waypoint, you can see your radial and distance from the VORTAC simply by pressing a "Check" button.

Foster Air Data Systems, Inc., specializes in RNAV equipment. The RNAV 511 has two-waypoint capability. In their literature, Foster claims, "The optimum number of waypoints for most RNAV operations is TWO. ONE generally is not enough and THREE or more can be too many...especially for single-pilot flight operations." I know a number of pilots, myself included, who would argue with this rationale—and it is interesting to note that Foster's higher-priced RNAV can store five waypoints! Let's put it this way: two-waypoint capability is perfectly acceptable, especially if the price is right.

The RNAV 511 is simple to operate, with a separate set of thumbwheels to enter radial and distance for each waypoint. Data can be entered to $1/10°$ and $1/10$ NM. Bearing and range information is displayed in digital form. For steering information on your HSI or CDI, a steering adapter option is required; this is available in either angular or linear format.

A unique feature called Range Monitor allows you to select either the Number 1 or Number 2 nav to drive the RNAV system. This lets you eat your cake and have it, too: use the RNAV with your HSI while navigating cross country, then, when you go to the localizer frequency, switch the RNAV to the Number 2 nav, where it will give you waypoints at the outer marker, runway threshold, or wherever else you desire.

A "Present Position" mode displays your bearing and range from the VORTAC in use.

Foster's fancier set is the RNAV 612, designed for medium piston and turbine aircraft. It has five-waypoint capability, and one of these waypoint settings can serve in an "Auto Waypoint" mode. This simplifies programming of waypoints along the enroute courseline. The unit also features the "Present Position" function, which it can store in memory as a waypoint, to facilitate returning to a given location.

At this writing, Narco is introducing two systems that will be similar in function to King's KNS 80 and KNS 81; so similar, in fact, that they bear the model designations NS 800 and NS 801.

Collins's Micro Line RNAV is the ANS-351, which has 8-waypoint capability. The companion IND-451 DME indicator shows, in addition to distance, your choice of groundspeed, ETE (estimated time enroute), or time-to-station.

Another feature is all-angle groundspeed, which displays accurate groundspeed up to 400 kn regardless of your direction of flight in relation to a waypoint. When used with the RNAV version of the Micro Line VIR-351 nav receiver, bearing to and from the waypoint is digitally displayed. There's also a "Present Position" check. CDI course deviation is linear.

Sperry offers the (formerly ARC) 400 Series RN-478A RNAV, with 3-waypoint capability. Waypoint data is entered and stored in two memory positions, and the third waypoint is set with the levers common to 400 Series equipment. Distance and groundspeed or time-to-station is displayed on the companion RTA-476A DME.

Loran. Even though Loran has been around since the 1940s, it is fairly new to aviation. However, it has caught on quickly, and at this writing, Loran equipment is the hottest seller in the entire avionics spectrum. The reason is the equipment—or some of it, at least—is relatively inexpensive and quite versatile.

The initial system, Loran A, was developed for maritime navigation during World War II. In the late 1950s, an improved system, Loran C, was put into effect. Loran traditionally has been operated by the U.S. Coast Guard.

Here's how Loran works: there are 17 chains, or groups of low-frequency transmitters, located worldwide. Each chain is made up of a master station and two to four secondary stations.

A master station broadcasts a series of nine pulses, followed by a series of eight pulses transmitted in turn by each of the secondary stations. The time difference between the sets of signals is measured, and by this means, lines of position are formed. The intersection of two or more lines of position establishes the aircraft's location, by latitude and longitude.

Using a typical Loran set, you select the nearest chain of stations and tell the unit where you want to go. With some equipment, you have to punch in the latitude and longitude of your destination. Other sets can file this information in memory, under a name such as the airport identifier (TEB for Teterboro) or perhaps your own designation (HOME, etc.). The set will then display a bearing and distance from its present position to the destination.

The Loran plots a great circle route, which is the shortest distance. You can, of course, insert waypoints for fuel stops, sightseeing, or to detour around unfriendly territory.

When you're airborne, the Loran will give your groundspeed and ETE. It can also provide steering information via its own built-in CDI or interface with your own CDI and even drive your autopilot.

Loran has limitations, some of which can be overcome by the more sophisticated equipment. At this point, I should mention that each make and model of Loran equipment has its own personality and capabilities, just like any computer and its software. A full account of the uses, features, and limitations of Loran equipment would fill a book. (I'm writing one.)

In general, Loran's limitations include propagation error, weather interference common to the low-frequency band, and inaccuracy due to station geometry. Also, there are gaps in the Loran coverage. Remember, Loran is operated by the U.S. Coast Guard, with the original sole purpose of serving the maritime interests, and there's very little of that in, say, Denver. As a result, coverage is weak in the so-called midcontinent gap, which takes in a considerable chunk of territory. This gap includes all or part of such states as Texas, Arizona, New Mexico, Colorado, Wyoming, Montana, and North Dakota. Sometime between 1987 and 1990 the FAA plans to install a chain that will provide coverage throughout the mid-continent gap.

Most of the airborne equipment is being provided by manufacturers of marine Lorans, who saw an opportunity in the aviation market that ironically was missed by the established avionics manufacturers. (King built a prototype in the early 1980s, then shelved the project.) Prices for light aircraft Lorans are in the $1500 to $4000 range. Some have been certified for enroute and terminal IFR, in most cases with a geographical restriction. At this time, the FAA is not certifying any Loran equipment for IFR approaches.

LORAN EQUIPMENT. II Morrow makes a line of very compact Loran navigators, 6.25 in. wide by only 2 in. high. All of their models are TSOd. The bottom-of-the-line Model 602 has an LED segmented display and can store 200 waypoints. Model 611 features an LED dot matrix display and automatic secondary selection, with capacity for 100 waypoints. Model 612 is identical to the 611, with the addition of a database, called FLYBRARY™, that has all airports with three-letter identifiers and all VORs stored in memory. An "Emergency Search" feature will provide nav information to whichever airport in the database is closest to the aircraft's position. Model 614 is a remote-mounted system that has VNAV (vertical navigation) capability and a function that can queue up 10 FLYBRARY airports or VORs in a flight plan sequence. At this writing, IFR certification is being sought for the 612 and 614.

ARNAV's Model R-21/NMS has an LCD dot matrix display and 200 waypoints. Based on the fact that it knows the aircraft's position and groundspeed, it can compute actual winds aloft, absolute fuel range and descent point (pseudo VNAV), with some pilot input required. An optional database includes virtually all public use airports and VORs, with a "Nearest Airport" feature that provides nav information to any one of the six airports that are closest to the airplane's position. A plain vanilla R-21, with less sophisticated software, is available at lower cost. The R-21 models are not IFR certified. For the pilot who wants an IFR box, ARNAV makes the R-40. It can interface with the Fueltron computer to provide fuel range computations without pilot input.

Narco's Model LRN 820 has an LCD dot matrix display with 200 waypoints, plus storage capacity for an additional 10 "present position" waypoints. The unit features automatic chain selection; the majority of general aviation Loran navigators require manual chain selection, after which they automatically select the secondary stations that make up the navigation triad. The LRN 820 is not TSOd or IFR certified.

Texas Instruments' IFR certified TI 9200, has an LCD dot matrix display and 99 waypoints. In addition, there is a database consisting of the 305 high-altitude VORs. The unit will store two flight plans with up to 20 waypoints each, plus reverse direction. The TI 9100 is IFR certified, has a gas discharge display and nine waypoints (it came out quite awhile ago). A more recent version, the TI 9100A, will store four flight plans, each containing up to nine user defined waypoints plus present position. There is also a remote-mounted version, the TI 91.

Micrologic's Model ML-6500 has an LCD segmented display and 125 waypoints. Its features include automatic chain selection and storage for nine flight plans with a total of 99 legs. The unit is not TSOd or IFR certified.

Nelco's Model AF 921 has a vacuum fluorescent display and 99 waypoints. Features include six flight plans of eight waypoints each, and dual waypoint tracking that provides continuous range and bearing information on one waypoint while navigating to another waypoint. The AF 921R is a remote-mounted version. Neither model is TSOd or IFR certified.

Offshore Navigation, Inc., manufactures the ONI-7000—the only Loran navigator that is currently certified for IFR enroute and terminal use in the entire U.S. national airspace; other Loran sets are restricted to VFR operation in certain fringe areas, such as the mid-continent gap. (Although no Loran unit is presently certified for IFR approaches, the FAA and a number of states are conducting research on the feasibility of non-precision Loran approaches.) The ONI-7000 has an incandescent display and 200 waypoints. Features include automatic chain selection and up to 99 flight plans with a total of 99 legs. The system utilizes up to four chains and eight stations simultaneously. It is remote mounted, and relatively expensive at over $14,000.

Foster's LNS 616 is more than a Loran navigator—it's an integrated nav system that uses both Loran and VORTAC input simultaneously. The computer compares the signals of both, determines which is more suitable for navigation, then uses the better signal as the primary source. The LNS 616 has its own built-in Loran receiver and uses the on-board VHF nav equipment. Its memory accommodates 100 waypoints, 26 flight plans, and 250 references. The cost of this system can run from over $14,000 to more than $19,000 with a database.

Weather Radar. There is no avionics equipment requiring as much pilot training as weather radar. Furthermore, using radar without training can get you in a lot of trouble.

I make these warning statements because, like so many of the other avionics we've talked about, weather-radar equipment has improved rapidly over the past few years, and its new capabilities are filtering down from the turbine world to the single-engine piston market.

But weather radar, however sophisticated, is an effective tool only when used by a knowledgeable pilot. The key word is *interpretation*, and that does not come easily. We'll explore some of the reasons for that, but first, let's see what radar does. (For brevity's sake, when I use the word "radar" all by itself, I'm referring to airborne weather radar of the type used in general aviation.)

Radar operates on the premise that a thunderstorm cell produces heavy amounts of rainfall. The function of radar is to measure that rainfall's intensity, paint a picture of its size and shape, and show you its azimuth and range from your plane. It does this by transmitting pulses of energy, which are reflected by the water droplets and returned to the receiver. The range of the return, or target, is determined by the time the round trip takes. If that time is 12.36 microseconds, the return is 1 mi away.

Radar will not paint clouds and is not likely to paint hail or dry snow, because their reflectivity is not great enough. What about icing? According to Bendix, radar might possibly detect intermittent moderate or heavy icing conditions associated with unstable air lifted by frontal action or mountain effects. In this situation, the cumulus cells are hidden by surrounding cloud layers, but could be spotted by radar.

Color radar uses chromatics to depict rainfall intensity. A red area indicates rainfall at the rate of 12 mm/hr (½ in./hr) or more. Yellow denotes 4-12 mm/hr, and green represents 0.1-4 mm/hr.

The problem with this type of reference is the pilot who is not trained in the interpretation of radar is likely to assume he will be safe if he avoids the red and yellow and flies into the green. After all, isn't that what the traffic signals have been telling us all these years?

But those visual cues can be totally misleading. The turbulence, which is what you want to avoid, will be occurring *somewhere* in that area, but it may be outside the rainfall. And according to studies by the National Severe Storms Lab, tornadoes are normally found in what would show up on the screen as a green area!

Then there's the matter of *attenuation*. This has to do with a characteristic of the X-band frequency range used by general aviation radar, in which the radio beam might be absorbed by rain that lies *between* the plane and a storm. Thus, your display could indicate that you have weather for only 20 mi, when in fact there is a bad storm further ahead, but the signal has been attenuated by the intervening rain.

This phenomenon was adjudged to be the cause of a DC-9 crash a few years ago. The crew thought they were going through the shallowest part of a storm when they were actually penetrating the worst of it. They were misled by attenuation.

Some radar manufacturers are now adding an attenuation-compensation circuit that turns up the gain in areas of moderate or high intensity rainfall to see behind that activity.

You also have to consider color in relation to range. A bad storm may show up as a green return at a distance of 150 mi, because the transmitter beam spreads out and becomes many miles wide at 150 mi. So if it hits something 150 mi out, only a tiny portion of the original energy that was sent out is going to hit that storm; this means only a little bit of that original energy is going to come back. The manufacturers try to cope with that by using a feature called STC (sensitivity time control) curves.

There's a difference in philosophy among some manufacturers in the use of manual gain control in the weather mapping mode. King allows you to adjust the gain, so if you have a large area of yellow and red, you can turn the gain down, and it will display only the highest levels of rainfall. They believe that you can interpret the conditions better with this technique, once you've experimented with the gain control in known weather conditions.

Sperry, on the other hand, does not have adjustable gain in the weather mode of their Primus 150. They feel that this model would probably be the pilot's first radar, and the system should be as foolproof as possible. They're concerned that with adjustable gain, a pilot might leave the control in its minimum position and unwittingly fly into some weather. Their more expensive systems have gain control. The theory is these sets would be operated by pilots who have had more experience with radar.

Another challenge to the radar user is to separate weather targets from ground targets. In some cases, you may have to tilt down to pick up a rainfall shaft, and if you are high enough, the antenna beam will also receive ground clutter. You must learn to adjust the tilt properly to determine which is which.

It should be clear by now that if you're going to use radar, you need to learn a lot about it. You can gain a certain amount of knowledge through experience, but if you want to survive the experience, use your radar for weather avoidance and *not* for penetration of thunderstorms.

Antennas. Until a few years ago, radar was strictly for multiengine aircraft, with the antenna mounted in the nose. Now there are a number of ways to install the radar antenna in a single-engine plane, some more effective than others. On the Cessna Centurion series, the antenna is housed in a pod suspended from the wing. This is somewhat unsightly, obstructs the visibility a bit, and adds some drag. (Fig. 11.4.)

There are pods mounted in the leading edge of the wing, creating a slight bulge. Cessna has them available for the current Skylane RGs. Then there's the Sperry (formerly RCA) WeatherScout I antenna that fits completely into the aircraft's leading edge. It's available as a factory installation on Mooneys and some of the Pipers. No unsightly bulge there, because the antenna is truncated into a banana shape. But this banana does not provide a free lunch, because when you truncate the antenna, you compromise the radar's capability.

According to Sperry, the tradeoff of having a truncated antenna

11.4. The Model P-210's antenna is in a pod suspended from the right wing. (Cessna Aircraft Co.)

that fits in the leading edge is a vertical beam width that is 15°, as compared to a full parabolic antenna that has a beam width of 8.5°–9°. This means the WeatherScout beam is a cone-shaped beam with a 15° vertical dimension. The more concentrated the beam, the higher energy level it has and the greater detection range of the radar, all other things being equal. An analogy would be a flashlight: the larger the reflector, the more intense and less diffused the beam. Likewise, the bigger the radar antenna, the greater the distance it will detect a storm.

The philosophy of the WeatherScout leading-edge system, says Sperry, is to give the pilot a radar that can provide 60–70 mi of detection capability to keep him out of the weather. (Light plane radars with full-size antennas generally offer ranges of 160–240 NM.) But according to radar expert Archie Trammell, small antennas in general—and the banana in particular—have more serious disadvantages than reduced range. Trammell also states they can distort the shape of a storm cell and not measure its true intensity.

Sperry's WeatherScout II and Primus 100 and 150 radars can be used in the full-size wing pods, as can the Bendix RDR-160.

A couple other gems of Trammell wisdom: a flat plate antenna will perform 15–25 percent better than a parabolic antenna. This is due in part to the side-lobe losses being much less with the flat plate antenna.

The performance of radar can be badly inhibited if its antenna is covered by a poorly made radome. Trammell likens this to using a powerful lamp with a dirty lens, and urges you to demand a written guarantee the antenna housing will have 85 percent transmissivity all across the radar window. A simple fiberglass shell will do the job if it's laid up properly. A honeycomb-structure radome is the surest way to go, but it is also the most costly.

One important feature that has filtered down from upper-end radars is pitch and roll stabilization. In straight and level flight, a radar beam sweeps in a 90° azimuth pattern ahead of the aircraft at whatever antenna tilt has been selected, to a maximum of $\pm 12°$. If the system is nonstabilized, the beam will point in whatever direction the plane's nose is pointing, pitching up or down, or banking in a turn. As you can imagine, a turn could thus produce ground clutter on one side of the screen and little or no return on the other side. A stabilized system is connected to the vertical gyro that normally provides sensing to the automatic pilot, keeping the radar beam stable while the plane is maneuvering.

King's KWX 56 digital color radar was the first moderately priced unit to offer stabilization (Fig. 11.5). However, it is not available for single-engine aircraft, because King does not believe that a suitable antenna has yet been developed for installation in singles.

A companion radar graphics unit, KGR 356, works with the KWX 56 and a version of the KNS 81 RNAV. With this combination, the screen can be made to show the weather picture, plus the location of the VORTAC in use, and all waypoints referenced to that frequency. If you find an enroute waypoint is too close to a storm cell, you can move the waypoint to any other location on the screen, using a joystick control. A track line will then give you navigational information to the new fix. The graphics unit will also display normal and emergency checklists. Other manufacturers offer graphics displays as well.

Actually, the radar equipment that was manufactured by King is now being produced under the Narco nameplate. The government required a divestiture when King was acquired by Bendix.

Bendix has a lightweight stabilized radar, the RDS-82, which *can* be installed in single-engine aircraft, using a wing-mounted pod. It presents a weather picture in four colors, highlighting in magenta the highest rainfall rate. A feature called Auto Pulse selects a combination of pulse repetition rate and pulse width most appropriate for each range of operation. The manufacturer claims that this results in dramatically improved resolution at every range.

At this writing, the lowest-cost color radar on the market (about $8000) and also the lightest (at less than 16 lb), is Sperry's Primus

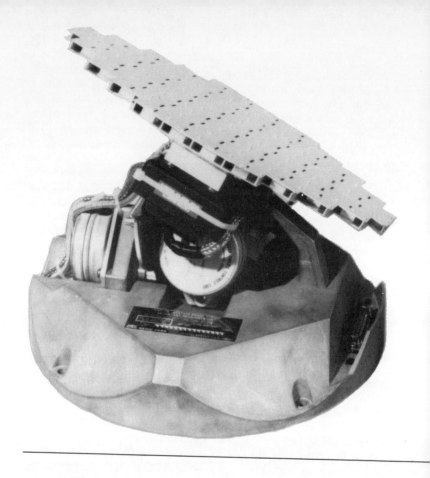

150. The set offers five range selections up to 180 NM. It does not have stabilization.

If you're wondering what happened to monochrome radar, it's still being offered by some manufacturers, but consumer demand is relatively low.

As I mentioned earlier, radar can detect objects on the ground. This is purely a nuisance when you're trying to map weather, but there is a ground-mapping mode that has a limited value for navigational purposes. Weather radar is not very good at this, because the long transmitter pulse required for weather penetration reduces resolution, and most terrain features are not picked up. Weather radar is at its best for ground mapping when distinguishing between surfaces that have sharply contrasting amounts of reflectivity, such as coastal areas or lakes.

Some radar sets use different colors for the ground-mapping

11.5. The King KWX 56 radar with flat plate stabilized antenna. (*King Radio Corp.*)

mode. One reason is to keep the passengers from getting excited if they were to see large red blobs appearing on the screen.

We can continue to expect the trickle-down effect to bring top-of-the-line features to the light plane market. For example, ground-clutter reduction is now available on the expensive models. And radar is being integrated with electronic HSI displays in the fascinating new EFIS (electronic flight instrumentation systems). This "glass cockpit" concept is now being used on jets and turboprops. We will probably see it on our piston-engine general aviation planes in five years or so.

Another thing that's on its way is the solid-state transmitter, which will replace the magnetron. This should add some reliability, although replacement cost will be higher. Down the road, the solid-state transmitter offers an interesting potential as an aid to detecting turbulence. Using Doppler technology, it may be able to measure frequency shifts in the returns. There's a theory that says if you're not getting frequency shifts, you won't encounter much turbulence. This concept is still in the experimental stage.

Now let's look at a weather avoidance system that takes another approach.

Stormscope. The Stormscope works on an entirely different principle from weather radar. To make a simple comparison, weather radar identifies rain, while the Stormscope identifies electrical discharges. Both phenomena are, of course, associated with thunderstorms.

The Stormscope was invented by Paul Ryan, an electrical engineer who got the idea following a white-knuckle ride in a thunderstorm while flying a Skylane with his wife and three children aboard. Here's how he described it to me:

> During that flight, I remember looking all around to see if any direction would be most inviting, but in all quadrants there were little streamers of electrical discharges, like thin ribbons of lightning. That prompted me to look into electrical discharge activity as a technological foundation, because I noted that when the turbulence was worst, the rain was not at its worst, but the electrical discharge activity was.

Ryan began a series of experiments based on the relationship between convective shear and the electrical discharges it creates. The violent updrafts and downdrafts associated with thunderstorms produce a separation of positive and negative electrical charges. Ryan correlated an increase in aircraft gust loads due to convective shear with this electrical activity.

He designed a weather mapping system comprised of four elements:

1. A low-profile antenna.
2. A receiver, tuned to a group of frequencies that center in the 50 kHz range.
3. A computer/processor that arranges the electrical image into a maplike fashion for display purposes. This processor also provides what is called a pseudo-range, that is, an estimate of the distance of the electrical discharges from the receiver.
4. A CRT display that shows the discharges in the form of dots, including azimuth and range information.

In 1976, the first Stormscope was brought to market. It was received enthusiastically by many pilots as a weather avoidance system that was less expensive than weather radar at the time and did not require radar's cumbersome antenna. Today's top-of-the-line Stormscope sells for more than the lowest-priced color radar, although installation cost of the radar will probably be higher. Perhaps more to the point, a number of users felt that the Stormscope did a better job than radar to help them avoid weather turbulence. Here are some of the advantages they found:

1. Weather radar scans an area ranging 90°–120° ahead of the airplane. Stormscope's top-of-the-line models display a 360° area, which is especially useful under circumstances where you want to keep an escape route in mind.

2. Radar, with its line-of-sight transmission, usually doesn't function well on the ground. Stormscope, utilizing long-wave frequencies, does not have this problem, enabling the pilot to plan his weather-avoidance route while on the ramp.

3. Radar is heavier, more complex, and more subject to malfunction than Stormscope.

4. The radar display requires considerable interpretation, as discussed earlier. Stormscope needs some interpretation, too, but it's more straightforward. (One experienced user puts it this way: "Don't fly into the dots.")

Radar has some advantages over Stormscope:

1. The radar display is more definitive. It can show the size and shape of a storm cell and, in the color units, it can present a graphic picture of the cell's intensity.

2. Radar's range accuracy is more precise.

3. Radar can be used for terrain mapping. With graphics capability, the screen can display navigation information, checklists, and other data.

All of the preceding reasons notwithstanding, the most important difference between radar and Stormscope is the difference in concept. Many people, including Stormscope's inventor Paul Ryan, feel that the ideal way to go is with both units on board—but that may be beyond the, shall we say, scope of some users.

When I asked one corporate pilot, who does fly with both, which he'd keep if he could have only one, he paused for a long time...and then declined to answer.

Stormscope, which is now owned by the 3M Corporation, offers the following models.

WX-10A. The CRT display has a 360° picture and a capacity of 256 memory dots. There are four range selections: 25, 50, 100, and 200 NM. The WX-10A is an improved version of the predecessor WX-10, and the improvement comes in the form of better definition. A thunderstorm can produce multiple discharges, which appear on the WX-10 as a series of dots streaming inward toward the aircraft. This "radial spread," as it's called, gives a somewhat ambiguous picture of the weather. The WX-10A has a new circuitry that plots the discharges more accurately, so that—paticularly in a high-intensity thunderstorm—the dots will appear more in clusters, and the radial spread will be reduced.

WX-11. This model has the added advantages of gyro stabilization. Let's see how this works in comparison with those models that do not have the feature. Suppose you are flying with a WX-10 (or WX-10A) and a thunderstorm is displayed directly ahead. If you do the sensible thing and make a heading change, the new weather information relative to your heading will be displayed, but the old information will also continue to appear until the computer removes it. This clearing operation could take up to 4 min—or until you press the "Clear" button.

The WX-11 (Fig. 11.6) uses input from a heading indicator with

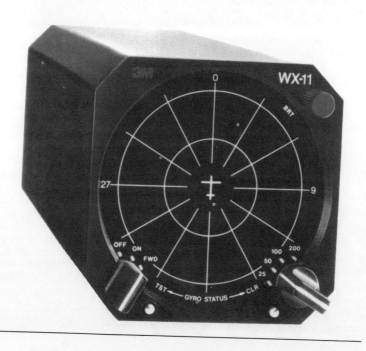

11.6. The Stormscope WX-11 features stabilization. (*3M Stormscope*)

a five-wire servo input, such as an HSI, to cause the display to "turn" with the airplane so you always see the weather information properly positioned in relation to the aircraft.

You can activate a gyro-status mode by pressing the "Test" and "Clear" buttons simultaneously. This will provide a display that tells you whether or not the gyro utilized by the Stormscope is operating. If the gyro is inoperative, the Stormscope display will be incorrect as long as it is connected to the gyro. However, by holding the "Test" and "Clear" buttons for 3 sec, you can bypass the gyro stabilization, and the WX-11 will then function like a WX-10A.

The older WX-10 can be converted either to a WX-10A or a WX-11.

WX-8. This is the economy model and is considerably different from the other Stormscopes. The display is not CRT, but liquid crystal. It covers an azimuth of 135° and is composed of 27 geographical segments using color to indicate range. Red segments represent activity in the 0-30 NM range; yellow is for the 30-60 NM range; green shows the 60-100 NM range. The computer and display are combined into one unit.

WX-12. Those people who want both Stormscope and weather radar can have the combination in a single display. The WX-12 does not have its own display but presents its information on the screen of most Sperry ColoRadar models. The pilot can opt for a picture of radar only, Stormscope only, or both simultaneously.

Stormscope has been accepted by the industry to the extent that a number of aircraft manufacturers now offer it as a factory option.

Airborne Telephones. An airborne phone gives you the ability to place calls to—or receive calls from—virtually any place that has telephone service. This type of equipment is purchased primarily for business purposes, but there are other uses as well.

Let's start by getting down to business. If you are winging your way to a business appointment, it's handy to be able to call the person you're meeting and give an updated ETA, especially if you're going to be late. And there's a pretty good chance that you'll impress your client with a call from 20,000 ft—unless he has his own airborne telephone.

You can receive calls from the office, if they know your flight plan and you've stuck to it pretty closely. You can call someone in another airplane, if he has a telephone and you know his location.

One of the radio-telephone manufacturers, Wulfsberg, publishes a Jepp-manual-size telephone directory, listing the phone numbers of air route traffic control centers, flight service stations, and airport control towers in case of an emergency when you can't reach them by radio. But the phone is not likely to be of service in the event of radio failure. Any panel that has a telephone will surely have dual nav/comm capability as well, and if the entire electrical system goes, it will take out the phone along with everything else. However, it is conceivable that you could lose your comm antenna, say, in icing conditions, and still have the phone antenna.

One use occurs to me that hasn't been promoted by the airborne phone manufacturers: if my destination airport does not have weather reporting, and the sequences for that area are questionable, I usually telephone the airport before departing to ask about the local conditions. It would be really nice to be able to do that from the air.

The airborne equipment consists of a UHF transceiver, antenna, and microphone and/or handset. You are required to have an FCC mobile radio telephone license. Prior to obtaining this, you must arrange for service with your local phone company and submit a letter of intent from them, along with your license application, to the FCC. Allow ample time for the FCC to process your application.

When you receive your license, you will be issued a "QM" number by the phone company. This is an identification number for billing purposes, and it is also encoded into your radio phone for receiving incoming calls.

Calls are placed and received through ground stations. At present, there are nearly 80 of these stations in the United States and Canada. Some of them are operated by telephone companies (referred to as Telcos), and others are independent companies called radio common carriers (RCCs). Since UHF is line-of-sight, your ability to communicate with a ground station will depend on your altitude.

Capabilities and operation methods will vary according to the equipment. I'll use the King KT 96 as an example, as it is the only unit currently built for the light-aircraft market. At 10,000 ft, the KT 96 has a range of about 110 mi, with the distance increasing to 220 mi at 40,000 ft.

There are 12 communication channels plus one signaling channel. You place a call in flight by contacting the nearest ground station on the channel(s) assigned to it. If you don't have this information, you can tune the band until you receive a strong dial tone. After getting your billing information, the ground station operator will put your call through, using telephone company land lines or microwave facilities.

If someone is calling you, the phone will ring through the cabin speaker. If you are too busy to answer the call, press the "Hang Up" button and the call will be canceled.

There are two charges for each completed call. The first one is levied by the ground station, for placing the call; the current rate is about $7.50 for the first 3 min and about $2.50 for each additional minute. The second cost is the normal telephone company toll charge from the location of the ground station to the location of the party you are calling.

If you've reached a Telco station, you can charge the call to your QM number. An RCC operator may ask for a credit card, third-party billing information, or they may insist on placing the call collect. The reason for this, they say, is that many of their bills go unpaid—and they, unlike the telephone company, cannot shut off your phone service for nonpayment.

As you can see, an airborne call is not cheap. But there's one consolation: there are no monthly charges. You pay strictly by the call, and, like conventional phone service, there is no charge if the call is not completed.

The KT 96 is a simplex system; that is, it works like your comm radio, in that you key the mike or handset to talk, cutting out the speaker as you do so. Most of the more expensive units are duplex, working like a conventional telephone.

The King unit can be connected through the KMA 24 audio console to the cabin speaker and microphone; it can be hooked up to the optional handset; or a dual installation can be made, where the handset functions like an extension phone. It is priced in the $2000 range.

Other radio telephone manufacturers are Astronautics, Fredrickson (now owned by Terra), and Wulfsberg. Their products are aimed at the turbine market and sell for upwards of twice the price of the KT 96.

Headsets. You can spend thousands of dollars for fancy comm radios and not be getting your money's worth if you're using an inferior microphone and/or speaker system. For a relatively small amount of money, you can upgrade with a good-quality mike, and a headphone will do wonders for improving your reception.

When a headphone and boom mike are combined, you have a headset. Broadly speaking, the headset world is divided between the noise-attenuating type, with muff-style earpieces, and the lightweight type, with an acoustic-tube mike and a single earpiece. If you want both noise attenuation and good communications in the cockpit, you will probably need an intercom.

There are four basic criteria you should use in selecting this type of equipment: clarity, comfort, convenience, and protection.

CLARITY. The ideal mike is one that will transmit voice modulation while picking up a minimum of cockpit noise and RFI (radio frequency interference). Noise-canceling mikes are specifically designed for the latter function, but all mikes must be positioned properly for effective transmission quality. More about this later. The headphone should enable you to hear what you want to hear, while blocking out ambient noise.

COMFORT. The degree of wearing comfort will depend on the weight of the unit, the design and padding of the headband, and the design of the earpieces. You may have to decide where your priorities lie, since the most comfortable headset may be the least effective in blocking out ambient noise.

CONVENIENCE. Most pilots would agree the handiest type of microphone is the boom mike with a PTT (push-to-talk) switch on the yoke. This is especially desirable when you're IFR, or even VFR in high-density areas, where you're likely to be kept busy flying, navigating, and communicating all at the same time. If your plane doesn't have a built-in PTT switch, you can get a portable one that fastens on the yoke with Velcro™. In headsets, compactness may be important, especially if the pilot does not leave it in the plane.

PROTECTION. FAA Advisory Circular 91-35 says, in part:

> The levels of sound associated with powered flight are high enough for general aviation pilots to be concerned about participating in continuous operations without some sort of personal hearing protection.
>
> Most long-time pilots have a mild loss of hearing. Many pilots report unusual amounts of fatigue after flights in particularly noisy aircraft. Many pilots have temporary losses of hearing sensitivity after flights; and many pilots have difficulty understanding transmission from the ground, especially during critical periods under full power, such as takeoff.

Noise-attenuating (muff-type) headphones provide an effective way of blocking out ambient noise, while at the same time permitting good communications reception. There is, of course, the alternative of using earplugs. While on that subject, I would urge you to give your passengers a break; if they're not wearing any other ear protection, pass out the packages of E-A-R® plugs.

Generally speaking, the lightweight headsets rate high in comfort and convenience while the muff-type headsets should excel in clarity of reception and hearing protection.

Microphones. Some microphones are designed to be noise canceling. A noise-canceling mike has ports on each side of a single diaphragm, designed so the sound pressure fronts arrive at both sides of the diaphragm simultaneously and in phase. Thus, the voice effect cancels out the ambient noise effect.

The tube-type mikes that are used in many lightweight headsets are not noise canceling. However, if the gain is kept low enough, you can achieve a good signal-to-noise ratio with this type of mike. Also, it is important to position the tube-type mike either at the corner of the mouth or just below the lower lip, so the sound of your breath does not cause sibilance when you are transmitting.

The following types of mikes are most commonly used in general aviation.

CARBON. This is the oldest and simplest mike design. Its frequency response and intelligibility ranks lowest, and it is bulky. Also, the carbon tends to pack after a while and must be shaken loose. The main benefit of the carbon mike is low cost.

DYNAMIC. This is probably the most widely used, due to its popularity with the manufacturers. It offers very good frequency response, intelligibility, and reliability. Dynamic mikes are generally in the medium price range.

ELECTRET. This mike has the reputation of producing the most effective noise canceling and the most intelligible voice transmission, although some experts will argue that a good dynamic mike is just as effective. An electret mike does have the undeniable virtue of being small and light. Some manufacturers charge a premium for the electret mike, although this is hard to justify on the basis of production costs.

Intercoms. These are excellent for instructional situations or other conditions when good communications are needed between pilot and copilot or passenger(s). An intercom helps save your voice, eliminates misunderstandings, and keeps the rear-seat passengers from feeling isolated.

Intercoms are available as built-in and portable models, the latter obviously being useful for the renter pilot or free-lance instructor. Entertainment systems—FM radio or tape—can also be integrated into some intercoms.

When evaluating an intercom, find out whether its failure in flight will affect your radio reception. Some units have the output of the comm radio connected through the input of the intercom amplifier. With this installation, if the intercom fails, you'll lose communication. There may be an override switch, but you still run the risk of not knowing immediately that you've lost your comm, and you might miss an important transmission. Conversely, if the intercom is connected in parallel with the comm set, you'll hear the radio even if the intercom is turned off or fails.

David Clark noise-attenuating headsets are designed to eliminate the high- and low-frequency noises damaging to hearing and to reduce the ambient noise level in the cockpit. There are two headset designs. One reduces the noise level by 24 db and the other by 27 db.

All units have the same ear seals and headpads of soft cushioning material for maximum comfort.

Clark offers three types of microphones, each with its own noise-canceling characteristics: two dynamic-mike designs and one electret mike.

The company also has a voice-activated panel-mounted intercom called ISOCOM. It fits into a standard $2\frac{1}{4}$ in. round instrument hole and can accommodate up to six headsets. The unit's circuitry cannot affect the modulation of the comm transmitter, as there are no active electronics between the mike and the transmitting radio.

Plantronics specializes in light-weight headsets and has two general aviation product lines: the MS 50 and the Starset. Both use the acoustic-type microphone.

The MS 50 is the original design Plantronics introduced 20 years ago. The earpiece and tube mike can be clipped to a lightweight headband that is provided, or it can be attached to the temple of the user's eyeglasses.

The Starset is designed to fit over the user's ear, and it has a telescopic tube arrangement that gives considerable flexibility in mike placement. In addition, its amplifier has an integrated circuit and features a two-position gain switch. This enables a copilot or right-seat passenger to reduce the volume without cutting down on the radio sensitivity.

Radio Systems Technology is the wire-it-yourself avionics kit manufacturer. The company offers two-station and four-station intercoms, with FM- or tape-player input and optional comm radio and cockpit voice-recorder interface equipment. Building time is estimated at 5–10 hr. The top-of-the-line model can be ordered prewired at additional cost.

Revere makes the HUSH-A-COM intercom. It can be used as a portable unit, operating either from a standard 9V battery or plugged into the aircraft's electrical system. Or the amplifier box can be mounted easily under the panel for permanent installation. The unit handles up to four headsets. A fail-safe system permits the pilot and copilot to transmit and receive when the HUSH-A-COM is in the "Off" position.

An optional cassette patch cord enables a pilot taking instruction to record all conversations between himself and the instructor—plus all radio transmissions—using a standard cassette recorder. Revere claims that this procedure can reduce learning time significantly.

Sigtronics makes portable and panel-mounted intercoms. They feature fail-safe circuitry that enables the pilot to use the comm radio even with the intercom off. The portable unit plugs into the aircraft's cigarette-lighter socket. The built-in unit can be mounted in the panel or below it with a mounting bracket.

Their STEREOCOM is a panel-mounted intercom that allows the pilot and passengers to listen to a stereo entertainment system while monitoring the VHF radio. The music is automatically interrupted when the VHF radio or intercom is active.

The company also offers headsets, including a stereo model.

Telex is probably the most familiar name in microphones, headphones, and headsets, having been in business since 1936. The company claims to offer the most complete selection of this equipment in the industry.

Telex also makes an intercom that can be permanently mounted or used as a portable unit. It is of standard radio width and features yellow LED indicators that come on when either the pilot or copilot uses the PTT switch. There is an input connection for a stereo entertainment system.

An Entertainment System. Here's an accessory you might want to use with your headphones and/or intercom strictly for pleasure. Avionics West, a California avionics shop, has put together an AM-FM cassette deck designed especially for aircraft use.

The EC-200 is packaged in a standard avionics configuration and plugs into a mounting tray like the other radios in your panel. The unit has a comm-interrupt feature. When this is activated, any comm transmission will override the music. It has a fast-attack, slow-decay squelch circuit, which means you won't miss the first syllable of a comm interrupt, but there's a one-second delay at the end of the speech, so the music won't pop in and out while Center is talking to you.

When the comm interrupt is active, everyone listening on the system hears the communication; it cannot be split so the passengers hear only the music. A switch will disable the interrupt if you don't want to hear the comm. On the other hand, if you get busy with communications and don't want to be distracted by music, you can activate a comm-only mode.

According to Avionics West, the FM tuner is specifically designed not to interfere with comm or nav equipment, whereas this is not true of the units made for automobiles—so if you install one of the latter, it must be placarded for VFR use only.

The EC-200 is on the options lists of Beech, Cessna, Piper, and Mooney, and is available through avionics shops. There is also a portable unit that fits into a case and can be plugged into a cigarette lighter. A suction-cup antenna goes onto the window. It has all the features of the built-in unit and is suitable for those who fly more than one plane. Two headphone jacks are supplied with the unit. Additional jacks can be purchased, and the EC-200 will drive at least a dozen headsets.

A few suggestions on selecting, using, and maintaining your equipment:

1. Choose the equipment based on the kind of flying you do. If you fly in an environment where radio transmissions are minimal, you probably don't need a fancy headset with PTT capability. If your plane is noisy, protect your hearing either with a noise-attenuating headphone or earplugs. If you're taking or giving flight instruction, an intercom is a worthwhile investment. A tape recorder interface is probably a good idea, too.

2. Try before you buy. This applies especially to the noise-attenuating headphones, some of which may seem okay for the first 5 min, but could become downright painful after a couple of hours. See if your friendly dealer will let you take the set on a trip, on approval. And get a radio check on the mike from a controller at least 20 mi out.

3. Care for your equipment. Keep it clean. Don't let your headset sit on the glare shield, where the sun will fry it. Coil the wires neatly. When pulling out a mike or headphone jack, hold the jack; don't yank it out by the cord.

4. Finally, carry an inexpensive spare mike and a minimum of one headphone. A dead mike or speaker could render you incommunicado, usually at the worst possible time.

Avionics Installation: Factory or Field? If you're in the enviable position of ordering a brand-new airplane, you can usually choose between getting the avionics installed at the factory or taking delivery on a "green" (unequipped) plane and having the boxes put in by the avionics shop of your choice.

There are arguments in favor of both factory and field installations, and we'll take a look at both sides. Even if you're not quite ready to buy a new airplane, don't flip ahead too many pages, because in the next section are some tips on selecting a radio shop for retrofit installations or service on your current bird.

THE CASE FOR FACTORY INSTALLATION. One benefit of buying a plane complete with radios installed is the bird is ready to fly when you take delivery and start making payments. Conversely, when you sign for a "green" airplane and then put it in the radio shop, you are "flooring" the investment and not getting the use out of it until the installation is complete. This factor is especially significant in the case of a plane that's expensive and/or a revenue producer. There's a possible exception to this: if your aircraft dealer installs the avionics, your payments may not begin until the finished plane is delivered.

Another advantage of factory installation is that the wiring is laid down before the interior goes in, which is theoretically neater and more efficient. The aircraft manufacturers take the position that they know best where components should be placed in their airframes to minimize exposure to heat and vibration.

Warranty service is another consideration. If one of those expensive black boxes becomes inop, the fault may lie either in the box itself or in the installation—for example, a cable shorting out somewhere in the bowels of the plane. If it is an installation glitch, and you're a thousand miles from home, how does it get fixed and paid for?

The answer varies somewhat, depending on the airframe manufacturer. If the factory installation is still under warranty, you can take your Beech or Cessna airplane to any authorized Beech or Cessna dealer, and he'll handle it. If it's a Piper, go to an authorized dealer for the brand of radios in the plane, and Piper will pay the dealer for fixing the faulty installation.

Are you out of luck if the radios were installed by a shop back home? Not necessarily. If the shop is a member of the AEA (Aircraft Electronics Association), an international trade organization, chances are you can take your plane to another AEA shop and the two outfits will work together to get you back on the road. Nonmember shops may also make a cooperative arrangement, but it's a good idea to ask the shop that does your installation how they would handle this kind of problem.

Financing is also something to think about. When your plane is equipped at the factory, your aircraft loan includes that equipment. If you're thinking of a field installation, better find out beforehand whether the add-ons can be covered in the financing contract. This can usually be accomplished if you get an avionics bid while negotiating the aircraft loan.

Here are the ways Beech, Cessna, and Piper perform their avionics installations:

For their executive and corporate aircraft (Bonanzas through King Airs), Beech installs radios in much the same way they build airplanes—on a highly customized basis. The company offers several radio packages but will also install anything else that's certifiable at a higher installation charge.

When the factory's avionics department receives the list you've spec'd out, they generate a computer drawing of a recommended cockpit layout. You go over this with your dealer and accept it or change it any way you want—as long as you leave basic instruments in the standard T formation.

A harness is built on a fixture that simulates your airplane model, and then all boxes and indicators are burned in for about 4 hr. The installation is tested prior to flight and again during the flight test. You get a copy of the wiring diagram, your dealer gets a copy, and a copy goes into the company files.

If you were to buy a preowned 1963 Bonanza, and it didn't have a wiring diagram, you could contact the factory and give them the serial number; they would access the salt mines in Hutchinson, Kansas, (honest!) where the older files are stored and send you a copy.

Beech claims to have testing equipment no shop could have—and possibly no other aircraft manufacturer, either. They bought some expensive gear several years ago in order to test their VLF (very low frequency) installations properly.

About 30 percent of Beeches, from the Bonanza on up, have standard avionics packages; the rest are customized, including all P-Barons, Dukes, and King Airs. Beech also offers classes and private instruction in the operation of avionics to customers who attend pilot ground schools at the factory.

During the many years Cessna owned ARC, that was about the only brand they would install in their aircraft. Now that they have sold ARC to Sperry, the customer may be given a broader selection of factory-installed radios from which to choose. Regardless of who makes the equipment, it's evaluated by the engineering department and, upon acceptance, a systems integration plan is drawn. The radios are thoroughly wrung out in a prototype airplane before going on the line. Each Cessna twin comes with its own wiring diagram. For the singles, installation manuals are kept by the dealers and may be purchased for $15-$20.

Piper offers packages of King, Collins, and Narco radios, with King being the largest seller by a wide margin. Custom installation packages may be ordered on some aircraft models; there's no great price penalty, but you'll probably have to wait longer to get your plane. Piper requires the avionics manufacturers to provide very stringent test data before they will accept any new equipment.

THE CASE FOR FIELD INSTALLATION. The greatest argument for installation by an avionics shop is flexibility. In theory at least, you can bring in any airplane and have whatever radios, clocks, tape decks, etc., installed just about anywhere you desire. Of course, you want to be sure your grand custom plan is feasible from an operational and maintenance standpoint. You also want to pick a shop that does good work and backs up its installations.

How do you find a good shop? Start by asking around. Talk to other airplane owners and to fixed-base operators who use independent shops. The FBOs can't afford to have their planes repeatedly grounded because of inept workmanship. What you're looking for is a shop that does neat, professional work; uses quality wiring, connectors, antennas, and other hardware; can make sensible recommendations as to equipment and panel layout; and will stand behind their work.

If all this is leading you to conclude the best shop is not necessarily the least expensive, you're absolutely right. Good work does not come cheap, and the lowest bidder may be cutting corners that can cost you time, money, and aggravation in the long run.

It's a good idea to get several bids and make sure you're comparing apples to apples. Get everything itemized in writing, including the hardware and the installation warranty. Some shops warrant their installations for the same period as the equipment warranty (usually one year), while others will stand behind their work for life.

Also, you should be given a good interconnect drawing. It should include a power-load analysis that indicates how much power the system draws and from what busses in the airplane. The wires should be properly marked. All this can save a lot of exploratory time, with the meter running, if something poops out in Peoria.

When you're evaluating a shop, take a look at some of their installations in progress. Even if you're not an avionics expert, you should get some sort of impression of the caliber of the technicians. Notice whether the wiring bundles are neat and appear as if they're going to lie in the plane properly, without any strain. Check the appearance of a finished panel. Would you be happy with it?

If you use your plane a lot, get as firm an estimate as possible on downtime. Allow for some optimism.

All other things being equal, you might want to give extra consideration to a shop that is a member of the Aircraft Electronics Association. This organization was formed in 1957 to help shop owners communicate better with each other and share knowledge about equipment capabilities, service procedures, and businesslike methods of operation. I have attended several AEA conventions and am impressed with the quality of their seminars and the interest shown by the members in keeping up with the ever-changing world of avionics. (Along with these comments, I feel I should state that I have a connection with AEA, in that I contribute articles to their magazine, *Avionics News*.)

Some shops contend an important advantage of a field installation is the independent shop's ability to obtain certain equipment, such as special-order HSIs and flight director displays, that is not on the aircraft manufacturers' option lists. Further, they say that even the everyday black boxes you get from the plane maker may either be too new—the first ones off the line without all the bugs eliminated—or old inventory that has since been improved. How's that for a no-win situation?

There is a growing tendency for the aircraft manufacturers to push for factory installations. Remember, avionics can account for a hefty 30 percent or more of a plane's selling price. The manufacturers understandably want to keep this business in-house, and will exercise their clout accordingly.

The avionics shops are concerned about this trend because they have investments of $100,000-$500,000 or more in test equipment, and with that kind of overhead, they claim they cannot survive on repair work alone. They need the profits from sales, and they predict as more of these profits go to the airframe manufacturers, the smaller shops will be driven out of business, and the customer will find it harder to get his radios fixed.

Your final decision about who should install your avionics will probably be based largely on your own temperament. Probably the easiest and quickest method is to buy a demonstrator already equipped the way you like it. The next simplest route is to order a factory installation. But if you're more of an individualist, and want to establish a good relationship with a shop you have confidence in, the custom installation could suit you better.

Some final hints:

1. Determine as well as possible the kind of equipment that will make you happy *before* you sign on the dotted line.
2. Talk to other airplane owners about their installations, both factory and field.
3. If possible, go to the AOPA (Aircraft Owners and Pilots Association), EAA, and other aviation conventions, where manufacturers display their latest products. This is a very volatile industry, and exciting new equipment is constantly coming out. Even if you don't get all the bells and whistles available, at least you'll be making an informed choice.
4. Know what warranties you're getting, both on the equipment and the installation—and get it in writing.
5. Go over the installation before you accept the airplane. Make sure everything is working, and check that there is proper provision for cooling and that all the components you can see are well secured, including the antennas.
6. Get checked out on the proper use of your equipment—including problem situations, such as in-flight failure of the digital display.

In sum, use the same rule for your avionics installation that you apply to everything else in aviation: don't take anything for granted!

Keeping Those Avionics Healthy. Whether you've got a panel full of oldies or are about to light up your life with a stack of digital radios, your goal is to keep them in the plane and out of the repair shop for as long as possible. Even if they're under warranty and you're not looking at $30-$50/hr plus parts, radio failure can range from annoying to hair-raising, depending on whether you're sitting on the ramp or plowing through solid IFR.

There are a number of things you can do to keep your avionics working regularly, and some of the most important steps should be taken when the radios are being installed.

In the previous section, I suggested you check for adequate cooling. This is something to look at in your present installation, even if you're not adding radios at this time. According to one avionics manufacturer, you can just about double the reliability of your radios with proper cooling. Very small changes in the cooling environment will provide very large payoffs. Therefore, some sort of cooling should be provided for any avionics stack with more than a couple of radios.

There are two ways of cooling the avionics in your panel. One is a ram air kit which uses a scoop to bring in outside air. A drawback to this system is the outside air may contain a lot of condensation, which is not good for radios.

A better way is to install a proper fan, and I use the word *proper* advisedly. It should have a minimum RFI motor. King markets one — the KA 20. As fans go, it is not cheap but is much less costly than the avionics stack it's protecting.

Unfortunately, when an airplane is sitting on the ramp in the summer sun, cockpit temperatures can build up that far exceed the TSO (technical standard order) specifications of the radios. So the radios are hot even before you turn them on. One suggestion under such circumstances is to get the doors and windows open while you're taxiing and do not turn the radios on until you remove as much of that heat as possible. You can get by with one comm while you're taxiing, and there's no need to use the DME — which is probably the hottest radio in your stack — until you're airborne. Heat shields are a wise investment, too.

A King spokesman offers the following thoughts on preserving the life of avionics and instruments:

> We're seeing a lot of our panel-mounted gyros coming back with contamination in them. When a vacuum pump fails, it can suck all kinds of trash back into the gyro. We're trying to determine what to recommend for that. And when you put a new hose on a vacuum system, there sometimes is

contamination in that hose that needs to be blown out.

When radios are installed in a stack, we think there ought to be a bit of space between them. The dust covers that our radios fit into actually have a lip which gives the radios about 1/8 in. of space. We've seen some that have been jammed together, and there needs to be room for air to get around the radios. Of course, heat rises, so you'll want to put the hottest radios on top.

We fly airplanes constantly here at the factory, and we have very few problems with the radios. We think it's important to turn everything on when you fly. An exception might be the radar, because the magnetron has a specific life, so you wouldn't want to run that when you're not actually using it. But for your other radios, we feel that regular use is better by far than letting them just sit there.

If your plane is not equipped with an avionics master switch, consider installing one to prevent inadvertently having a radio on when you fire up the engine. This can cause problems, because the starter motor is a big inductor, and when voltage is applied to it, it draws a lot of current over a short period of time, and that can induce a voltage spike. There are pushbutton avionics master switches on the market that light up to call your attention to the fact that they've been left on after the battery master has been turned off. You should also have a backup switch, so you don't lose your radios if the primary master switch fails.

Cigarette smoke can do to your gyro filters what it does to the cigarette filters. Even if you don't want to believe that smoking will shorten *your* life, you'd better believe it will shorten the life of your gyros. If you must smoke or allow others to do so in your plane, check the filters often and replace them as necessary.

These steps will involve a little trouble and some cost, but they'll pay off in helping to keep those valuable avionics doing their job when you need them.

12. Instruments

A blinking EGT gauge... What the fuel computers won't tell you... More accurate airspeed control... An instrument that can save you money on navcomms.

THERE are a number of highly informative instruments that may not be on your plane's panel. They can reduce pilot workload and add an extra measure of safety and efficiency to flying. Let's look at some of them.

EGT/CHT Gauges. You'll recall in Chapter 4 I discussed the importance of proper leaning and the virtual necessity of having an EGT (exhaust gas temperature) gauge for this purpose. Many planes come equipped with a single-probe instrument. This is better than nothing, but it has an inherent weakness: the single-probe system may not always be reading the temperature of the leanest cylinder, since no one cylinder is the leanest at all times. This mitigates against leaning to the indicated peak EGT, because that may bring one of the cylinders to the *lean* side of peak. So pilots using this system tend to play it safe, fly on the rich side of peak, and thus lose the fuel efficiency they would enjoy by operating at peak at approved power settings.

A multiprobe system not only helps identify the leanest cylinder, it also indicates the spread, or EGT variation, from one cylinder to another. Alcor asserts that by knowing each cylinder's EGT, you can reduce the spread and thus increase efficiency, by "fine-tuning" your engine in flight through the judicious use of carburetor heat, throttle, or alternate air.

Therefore, if you have a single-probe system, you should consider upgrading to an instrument that can show the exhaust gas temperatures of *all* cylinders—at least one at a time or, better yet, simultaneously.

Here are several of the interesting new crop of EGT gauges, one of which is even computerized. Each of the following manufacturers uses a different technological approach, offering you a choice that goes beyond mere brand-name preference.

ALCOR. This company has more experience than any other in the design and manufacture of EGT gauges. Alcor's first product was a single-probe EGT gauge, which has been factory-installed on many production aircraft. Some years ago, they introduced the Combustion Analyzer. This is a multiprobe, single-reading system; it has a probe on each exhaust stack and gives a temperature reading on one cylinder at a time via a control knob. They also make a fancier version that indicates both EGT and CHT (cylinder head temperature), one cylinder at a time.

Now Alcor has overcome the single-reading disadvantage with a multichannel combustion analyzer that shows the EGT of all cylinders simultaneously (Fig. 12.1). It has an analog presentation, with increments of 25°F, just like their previous EGT meters. A multichannel CHT gauge is also available.

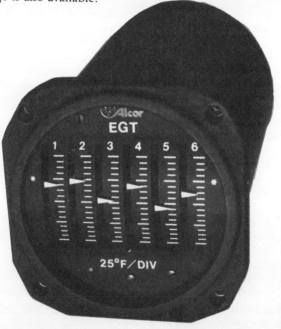

12.1. The Alcor Multi-Channel Combustion Analyzer. (*Alcor, Inc.*)

INSIGHT. This company makes the Graphic Engine Monitor, or GEM, which uses computer technology and a gas discharge display.

The GEM has the distinct advantage of monitoring both the EGT and CHT of all cylinders simultaneously. Its display is in the form of electronic light bars. These bars light vertically to indicate EGT, and one bar blanks out in each column to show that cylinder's CHT. To establish the leanest cylinder during cruise, simply lean the mixture until one of the bar columns begins to blink. This signifies that that cylinder has just gone slightly lean of peak. Press a reset button, which puts the GEM in the monitor mode, and richen the mixture by one bar (25°F) to arrive at peak EGT. With the instrument in the monitor mode, the computer puts the temperatures of each cylinder into memory. If any cylinder's EGT rises by 50°F, that column will blink, alerting the pilot to the situation.

One obvious extra benefit of this system: if you descend from cruising altitude and forget to enrich the mixture, the instrument will start blinking as soon as you bring the power back in.

Two meters are necessary for twin-engine aircraft. An added benefit for twin operation is the ready identification of the dead engine in an engine-out situation: dead light bars = dead engine.

Figures 12.2–12.6 illustrate various ways to use the Graphic Engine Monitor to detect and diagnose malfunctions, according to the GEM's manufacturer.

12.2.

INDICATION: Gradual or sudden rise in the EGT of one cylinder. Display will blink to warn you when this occurs.
PROBABLE CAUSE: Fouled or defective spark plug or faulty ignition wire.
WHAT TO DO: Switch to left and right mags momentarily to determine which plug or lead is at fault. Faulty plug will cause EGT to drop; all other cylinders will rise. The plug may clear itself when mags are switched.
ON THE GROUND: Check plug and ignition system; repair as necessary.

12.3.

INDICATION: Above-normal CHT of one or more cylinders.
PROBABLE CAUSE: Faulty engine baffles or cooling obstruction, such as a bird's nest.
WHAT TO DO: Open cowl flaps and reduce power. If temperature is dangerously high, land as soon as possible.
ON THE GROUND: Inspect engine baffles, and check for cooling obstructions.

12.4.

INDICATION: Rise in EGT of all cylinders. All bars on the display will blink.
PROBABLE CAUSE: Faulty magneto.
WHAT TO DO: Check that the mag switch is on "BOTH." Enrich mixture to return EGT to normal. Mag failure is a serious problem and must be treated accordingly.
ON THE GROUND: Repair ignition system.

12.5.

INDICATION: Abnormally low EGT.
PROBABLE CAUSE: Intake valve not opening completely; low compression. Could result from exhaust-system leak, cracked pipe, loose or cracked flange, defective gasket.
ON THE GROUND: Check compression, and inspect exhaust system.

12.6.

INDICATION: Decline in EGT uniformity; most noticeable at cruise power settings.
PROBABLE CAUSE: Fuel-system restriction or dirty injection nozzles.
ON THE GROUND: Clean fuel nozzles, and check fuel system if symptoms persist. (*Insight Instrument Corp.*)

ELECTRONICS INTERNATIONAL. This company makes a line of digital EGT and CHT gauges that have LCD displays. They show precise temperature readings to within one degree, as opposed to the Alcor and Insight units, which display relative readings rather than actual temperatures. The company claims precision leaning with their digital gauges can save ½–1 gph over the use of competing instruments.

None of the EI gauges displays more than one temperature at a time, but all are multipurpose units. For example, the gauge shown in Fig. 12.7 monitors a single channel of EGT and one channel of CHT, with a switch for selecting EGT or CHT. Moving up the ladder, there's an EGT analyzer that can be switched to monitor each cylinder, one at a time, and a CHT analyzer of like capability.

12.7. The Electronics International EGT/CHT gauge. (*Electronics International, Inc.*)

Fuel Computers. When you're up there boring a long cross-country hole in the sky, there are three questions you like to have ready answers for:

1. How much fuel do I have left?
2. At what rate am I burning my fuel?
3. Based on 1 and 2, how long can I keep flying?

This is another way of asking that very important question: "Am I going to get there—with fuel to spare?"

Thanks to microprocessors, which have brought us so many other good things, there are a number of instruments now on the market that can help provide the answers to those questions, with a high degree of accuracy... *as long as they are used correctly.*

It's important to understand what they will do and what they will not do.

The simplest of these products is a totalizer. It keeps a running tab on how much fuel has been consumed from startup to the present moment, leaving the rest of the mathematics to you.

The next step up is a totalizer and fuel flow meter. It tells how much fuel you've used and the rate at which you are using the fuel that is left.

From there we progress to an instrument that advises you of the fuel consumed, fuel remaining, fuel flow, and time remaining at your present power setting. It will also warn you when fuel is running low.

Let me emphasize that none of the instruments we are assessing actually *measures* the fuel in the tanks. In short, even though they may seem to function as fuel gauges, they are *not* fuel gauges and should never be regarded as such.

Here's how a fuel computer typically works: you program in the amount of usable fuel on board. The instrument then measures the fuel flowing to the carburetor or fuel injector by means of a transducer, which has a paddle-wheel rotor whose movement is sensed by an infrared light beam. This movement is translated into fuel flow by the computer, which then keeps a running tab on how much fuel you've used and how much you should have left, *based on how much fuel you told it you had initially.*

The point I hope I'm making is that if you do not enter the correct amount of fuel on board, you will not get a correct readout of fuel used, fuel remaining, or time remaining. As the computer folks are so fond of saying, "garbage in, garbage out."

"Well," you say, "what's the big deal? I know my plane's usable fuel. All I have to do is top my tanks and program that amount into the computer."

That's true in theory, but it is subject to possible error. For example, if the lineperson leaves a mere ¼ in. space at the top of a long-span tank, you could have up to 2 gal less per tank than the usable-fuel quantity you programmed into the computer. In itself, this is hardly likely to be a life-and-death situation—surely you won't be flight-planning that closely—but this type of error can multiply over a period of recurrent fuelings.

A similar situation can arise if your plane is not level when it is fueled. If the plane has a complex system of interconnected fuel tanks, air can be trapped in the system, again resulting in your having less fuel on board than you thought you had.

Another consideration: the transducers, which are not made by the instrument manufacturers, vary in size, and the computers must be calibrated to them. This introduces an additional possibility of error, especially when the transducer is replaced in the field.

Finally, there is an error that can result from temperature changes. For every 60°F of temperature change in your fuel load, there's a 3 percent error in the volume. For example, suppose you gas up at Tucson, where the tanks have been baked by the sun to 120°F. You take off, climb high and cruise where the OAT (outside air temperature) is 0°F, and there the fuel will contract 6 percent.

That gas is denser by 6 percent, so you may get the same mileage, but the computer is measuring volume, not density. Thus you will get a discrepancy between the gallons used, as shown on the instrument, and the gallons it will take to refill your tanks. One instrument manufacturer tells me they have received many calls from customers who don't understand this principle.

Here are some of the manufacturers of fuel computers and their products.

Shadin makes the Digiflo™ instrument, which fits in a standard instrument hole and has an LED display. The Digiflo shows fuel used, fuel remaining, and time remaining. The "time remaining" display flashes continuously when the reading drops below 30 min.

The instrument has a "keep-alive" memory circuit, powered by a nicad battery that is trickle-charged by the aircraft's electrical system. There are several models, for singles and twins, with fuel readings in gallons or pounds. On twin-engine installations, the gauge will immediately identify a failed engine.

SDI/Hoskins offers a choice of instrument shapes. The CFS 1000A fits a standard 3-in. hole and the flatpack CFS 1001A is compatible with a standard avionics-width stack (Fig. 12.8). The CFS 2000A and 2001A are twin-engine versions.

12.8. The SDI/Hoskins fuel computer. (*SDI/Hoskins*)

These models have a constant display of fuel flow, plus selectable displays of fuel used, fuel remaining, or time remaining, in addition to a timer function. Fuel quantities are displayed in gallons or pounds, at the pilot's option. There is a low-fuel annunciation, a low-standby battery warning, and a fill-up mode. The latter feature enables you to preprogram your airplane's usable fuel quantity into the computer's memory. Then, whenever you top the tanks, you can press the fill-up button, and that gallonage will be entered automatically. Again, you want to be sure you actually have all those gallons in your tanks.

Silver—now owned by ARNAV, the Loran manufacturer—has two model lines for piston-engine aircraft. They utilize incandescent displays.

The Fuelgard is a totalizer and flow meter, displaying either fuel used or fuel flow, as selected by the pilot. The instrument is a compact 1 in. × 2.9 in.

The Fueltron® fits in a standard instrument hole and shows fuel flow, plus selectable displays of fuel remaining, fuel burned, or time remaining. The single-engine model is calibrated in gallons, while the twin-engine model is calibrated in pounds.

Alcor's Tru-Flow uses an LCD on an instrument that is 1 in. × 3 in. Fuel flow and total fuel used are displayed simultaneously in the single-engine model, and the same-size unit is usable for twins.

The Tru-Flow can be configured by the owner to display in U.S. gallons, Imperial gallons, liters, or pounds. It can also be custom-programmed; for example, the flow reading can be averaged to provide a stable indication of unstable fuel flow.

If you are considering any digital fuel computer, bear some of these thoughts in mind:

1. Have it installed by someone who has done it before. Some owners who gave the work to first-time installers paid for their learning curve at more than double the time estimated by the manufacturers.

2. Don't be lulled into a false sense of security by those precise digital readouts, especially regarding fuel quantity and time remaining. Remember, these units do not measure fuel. Verify their indications with the customary visual preflight checks, time checks, and cross-references to the fuel gauges.

3. Make allowances for a gas heater or other accessory that burns fuel from the main fuel supply. This fuel consumption will not be computed by the instrument.

Digital indicators may tempt pilots of twins to lean the engines to obtain matching fuel-flow readouts. The correct procedure is to lean each engine individually by EGT/TIT indications. The respective fuel flow readouts should be close to one another. If they are not, there is probably a malfunction in one of the engines, such as a leaky manifold, injection or carburetor problem, etc.

As an example, the owner of one fuel computer was seeing a large fuel flow discrepancy between the two engines of his Cessna Skymaster. He finally discovered the problem was caused by one of the tachometers, which was 100 RPM off. Moral: do not lean slavishly to achieve the expected fuel flow, but instead, refer to the EGT/TIT as primary gauges, and use the digital fuel flow for fine-tuning and as an aid to discovering malfunctions. Special care must be given to monitoring turbocharged engines, which achieve higher temperatures than normally aspirated powerplants and are less stable.

A digital fuel computer is a useful device that will serve you well, provided you use it properly.

Speed Control Instruments. Your airspeed indicator does a pretty fair job of telling you the indicated airspeed in level flight. It becomes less accurate as an approach-speed indicator, largely because the pitot tube is then moving at an angle to the relative wind. Also, it lags in showing airspeed variations due to attitude changes caused by gusts or control movements. The correct indicated approach speed will vary according to aircraft weight.

Nor can the ASI be relied on to warn of an impending stall, due to the preceding reasons, plus the fact that the stall speed increases as the angle of bank increases. For example, the published flaps-up stall speed of a Skyhawk, loaded to its maximum aft CG, is 39 KIAS (knots indicated airspeed) with wings level and 56 KIAS at a 60° angle of bank. At maximum forward CG, the readings are 5-6 kn higher. And, of course, these speeds change when flaps are deployed.

However, there are speed-control instruments designed to help you maintain the desired airspeeds for approach, climbout, and stall avoidance—regardless of attitude, flap setting, and loading. In addition to providing increased safety—especially in short field and/or high-density altitude conditions—these instruments have the potential for saving wear and tear on brakes, tires, and nerves; many pilots routinely make their approaches faster than they need to, simply because they are justifiably reluctant to rely on the ASI.

A speed-control instrument has the additional value of providing some backup information in the event of pitot system blockage or other malfunction of the ASI.

Here are two companies that make these instruments.

SAFE FLIGHT. Leonard M. Greene, founder of Safe Flight, invented the stall-warning indicator found in one form or another on most civil and military aircraft. The company's SC-150 is an angle-of-attack system that combines a lift sensor with a computer and indicator. By measuring and displaying changes in angle of attack, the SC-150 helps the pilot to maintain precisely the proper speeds for best climb angle, downwind and base leg, normal approach, or short field approach.

The display is customarily mounted on top of the glare shield, allowing the pilot to keep the runway in sight while monitoring the instrument.

CONTROLLED FLIGHT MECHANISMS. Morgan Huntington, an engineer, has invented a speed-control device that works on a different principle than the Safe Flight instrument. Huntington's Lift Reserve Indicator does not have a computer. Its sensor is a probe that projects from the underside of the plane's wing at an angle of about 50°. The probe has two holes, one on each of two surfaces that face the relative wind. At different angles of attack, the air pressure is, of course, different on each hole. A meter measures and displays the differential pressure. The instrument uses as its point of reference what Huntington calls POWL-zero (potential of wing lift). The POWL-zero, therefore, is the point at which no surplus kinetic energy is available, and it signifies the onset of mushing sink, the precursor of the stall.

The Lift Reserve Indicator has been given intensive scrutiny by the AOPA Air Safety Foundation and Cessna's engineering department. I talked to representatives of both organizations, and they were highly enthusiastic about the instrument.

Assorted Instruments. There's an unending variety of instruments popping out of mail-order catalog pages and the ad columns of aviation publications. I've selected a few instruments made by Davtron, as examples of some of the more interesting products available.

If you have a mechanically tuned ADF, you might be interested in Davtron's Model 701B digital ADF indicator. It's a crystal-referenced frequency counter that converts any nondigital ADF to digital operation. This will enable you to tune in a station before you're in range to receive it, and you'll enjoy the dual benefits of dial scanning and digital accuracy.

Their Model 903 ID digital VOR indicator provides a digital bearing to or from the VOR in use (Fig. 12.9). When a localizer is tuned in, the instrument utilizes a bar and numerals to indicate the direction and degree of deviation from centerline. In addition, the VOR or LOC identifier is flashed one letter at a time.

A digital OAT gauge can give pleasant relief from the neck-twisting and squinting often necessary to read the standard analog

INSTRUMENTS

12.9. The Davtron Model 903 digital VOR indicator. (*Davtron, Inc.*)

OAT instrument. The digital display is less likely to be misread and is more likely to alert you to significant temperature changes. Davtron's Model 301C is calibrated in Celsius, Model 301F in Fahrenheit. Model 655-2 displays five functions on one instrument: OAT Fahrenheit, OAT Celsius, pressure altitude, density altitude, and aircraft voltage.

Davtron also offers the obligatory digital clocks. Model 811B is a three-function unit, serving as a clock, a flight time recorder, and an elapsed-time meter. It has an incandescent display. Model 800 has a liquid crystal display and shows local time, GMT, and elapsed time with count-up and countdown, plus alarm.

There are, of course, other good manufacturers turning out other good instruments. Perhaps this brief appetizer will tempt you into looking at the complete menu.

13. Autopilots

Rate-based vs attitude-based... The building-block concept... What an HSI does for you... The flight director: pro and con.

PEOPLE become pilots for many reasons, but I think it would be safe to say most people who fly enjoy the hands-on activity of controlling the airplane. Why, then, buy an electro-mechanical device that does the flying for you?

For a number of reasons. If you fly single-pilot IFR, you know how busy you can get—copying amended clearances, trying to keep track of your position while being vectored hither and yon, etc. Even if you're strictly a VFR pilot, coming into a TCA (terminal control area) can give you an IFR-like workout, and being able to dial in the headings and altitudes can help keep you ahead of the game. On a long trip, getting spelled at the wheel can help stave off fatigue. And if the unexpected pops up, such as weather or a malfunction, it's extremely helpful to get the piloting taken care of while you assess your options.

Finally, if you should ever get into a situation where you become spatially disoriented and realize it, an autopilot might save the day. (But that should not be a rationale for pushing into conditions you're not qualified to handle!)

Autopilots range in price from a simple wing-leveler at less than $3000 to a three-axis, integrated flight-control system, complete with flight director, in the $20,000 range. Altitude and vertical speed preselect can add anywhere from $2000 to $10,000 to the tab. And don't think the full-house system is reserved for cabin class. Jim Irwin, president of S-TEC Corporation, has this to say on the subject:

> A few years ago, it would have been very difficult, if not impossible, to sell a full two-axis system to the owner of a Skyhawk or Archer. But in recent times, that seems to have changed considerably. We've sold a number of *full flight director systems* on Skyhawks.

AUTOPILOTS

There has been some semantic disagreement on what is a two-axis system and what is a three-axis system. Most people in the industry agree that the defining factor should be the number of axes an autopilot controls with servos. Usually, a single-axis autopilot has servos only on the ailerons; a two-axis autopilot has servos on the ailerons and elevator; a three-axis autopilot has the addition of a yaw damper, with servos on the rudder.

Some years ago, Mooney, with safety in mind, installed a wing leveler as standard equipment on all their aircraft. The system was called Positive Control, and it certainly was. It was "on" all the time, unless you disengaged it by continually pressing a painfully small spring-loaded yoke button. The well-intentioned PC was not popular with the customers, and finally passed from the scene.

Position-based and Rate-based Systems. Autopilots are said to be either *position based* or *rate based*, depending on whether they get their sensing information from an attitude gyro or a turn coordinator. The two designs perform rather differently from one another, particularly in the lateral, or roll, axis. Let's analyze how each type operates, using the autopilots of S-TEC and King as examples.

The S-TEC autopilots are rate based. Their systems use a turn coordinator in roll as the basic azimuth-rate sensing device. They differ from an attitude-gyro system in that they are calibrated to provide a maximum *rate of turn*, as opposed to a maximum *angle of bank*.

In the case of S-TEC, the maximum is set at 3°/sec, the standard rate turn. If you command a heading change of 15° or more, the autopilot will command an initial turn at the full rate, and as the plane gets within 15° of the desired heading, the autopilot will gradually shallow the bank angle until the heading is reached, at which point the autopilot should produce zero bank angle and zero turn rate.

If you move the heading bug less than that, say 5°, the commanded turn will be at a proportionately lower turn rate.

Of course, if the rate of turn is fixed, the *bank angle* will vary with the airspeed. So if a plane has been slowed to, say, two-thirds of its cruising speed for intercepting a localizer, its angle of bank will be shallowed proportionately under the S-TEC system. The autopilot manufacturer feels that this is an advantage because there is less roll to remove during the intercept than you would have with a system that maintains the same maximum bank angle regardless of the airplane's speed.

Now let's look at the opposition.

The King systems are position based. Roll and pitch attitude are controlled by a vertical gyro, often called an attitude gyro, which has a roll and pitch attitude signal. Any change in the attitude you're flying is sensed in the computer by the signal from the vertical gyro. The computer then processes a signal which is sent to the servo actuators, which move the controls to return the airplane to its original position.

For example, suppose you're flying a Mooney 201 equipped with a King KFC 150 autopilot. You're heading north, and you crank in a course change of 60° to the left. The autopilot commands a turn to the left, and the aircraft rolls in that direction at the rate of about 5°/sec until it reaches its maximum bank angle, which is set for that plane at 20°.

As the airplane turns to within 20° of the new heading, the autopilot begins to level the wings at the rate of 1° for each degree of heading change. When the plane reaches a heading of about 3° from the new course, the system will slow the unbanking process by about one-half the normal rate, to avoid overshoot.

The maximum angle of bank will be the same, regardless of the plane's airspeed. Therefore, when the plane is slowed, the rate of turn will increase. King feels this is an advantage, because at slower speeds, you are typically coming into an approach situation—perhaps intercepting a localizer, whose beam width is one-quarter that of a VOR. Therefore, when capturing the localizer at some cut angle, you want to turn quickly if you want to avoid overshooting.

Well, now you've heard both sides of the story, and it may suggest something to look at when you're getting ready to buy an autopilot. I'll have a few other thoughts on that subject later, but right now, here are some of the current autopilots.

CENTURY FLIGHT SYSTEMS. Century has been known by a variety of names, starting with Mitchell Industries and including Edo and Edo-Aire. In 1983, Edo decided to get out of the autopilot-manufacturing business, and three of their key employees bought the division, which was given yet another name, Century Flight Systems.

The current owners are continuing to manufacture two generations of autopilots. The older series, Century I, II, III, and IV, is still being turned out to accommodate the many aircraft for which the units are given supplemental type certificates.

The more modern line consists of the Century 21, 31, and 41. This line offers improved lateral radial coupling. Also, the glideslope performance of the 31 and 41 is better (according to the manufacturer), in that it anticipates coupling without pitchover. These autopilots are capable of capturing the glideslope from above.

The 21 and 41 were designed at the same time and behave in much the same way. The 21 is single-axis, and the 41 is two-axis.

AUTOPILOTS

147

When the DG (directional gyro) is providing heading information, if there is full-scale CDI needle deflection, the 21, 31, and 41 will always go into a 45° intercept. The 21 and 41 use a combination of course error and needle deflection to determine whether to make a more modest intercept. The 31 makes its decision by needle deflection alone.

All three systems have a maximum bank rate of 5°/sec. The bank angle starts to shallow when the azimuth error (heading change) is reduced to 32°.

The autopilots will go into a soft mode when the aircraft is on track. The amount of softening increases with time in the 31, achieving full-soft in 70 sec. Under certain conditions, the 21 and 41 will get full softening immediately. In the soft mode, the roll is limited to 8° and the activity is desensitized, to minimize needle chasing in case of VOR scalloping or a big deflection when passing over the station.

The 21 and 31 are "one-box" systems, in that the controller, computer, and annunciator are all packaged together. This helps keep costs down and simplifies installation.

The two-axis Century 31, which came out about two years after the 21 and 41, was given some logic simplifications, in an effort to duplicate most of the functions of the 41 in less than half the space. The Century 41 has separate boxes for the controller, computer, and annunciator. One advantage of the remote annunciator is you have greater flexibility in locating it on the panel for optimum visibility. The main reason for spending about $3000 more for the 41 than the 31 would be to get a flight director, which isn't available on the 31.

All Century 41 systems are approved with coupled go-arounds. Some of the aircraft manufacturers require that the autopilot disengage when the go-around mode is entered. At this writing, you can get the coupled go-around if the 41 is installed by a radio shop, but that may change; Century is concerned about having consistency between factory and field installations, so as not to confuse the pilot who flies a variety of airplanes. Altitude preselect is not available for either the 31 or 41.

KING RADIO CORPORATION. King, which undoubtedly has the most varied avionics output of any manufacturer, makes several lines of autopilots. For years, the 200 series has been the bread-and-butter offering. Recently, however, King introduced the 100/150 series, which boasts digital circuitry. This results in a smaller parts count, permitting a one-box configuration.

The new autopilots have an interesting feature that may confound first-time users who fail to read the operating manual. There is a preflight test button that, when pressed, starts a sophisticated test sequence that turns on all annunciator lights and checks the essential electronic elements of the system. The autopilot will not engage unless the test is initiated and passed.

The KAP 100 is a single-axis autopilot. It is capable of automatic interception and tracking of VOR and LOC. The KAP 150 is a two-axis autopilot. Both the KAP 100 and the KAP 150 come with attitude gyro and DG; an HSI may be substituted for the DG at extra cost.

The KFC 150 has the same capability of the KAP 150, with the addition of an HSI and flight director. The flight director has a single-cue display. There is no go-around mode on this series. A yaw damper and altitude/vertical speed preselect are available as options.

The older KAP 200 and KFC 200 are less compact, heavier, and more expensive. However, you may not be able to choose between the 200 series and the 100/150 series; it really depends on which of the models has been certified for your airplane.

S-TEC CORPORATION. S-TEC is a relative newcomer to the autopilot scene. The company was formed in 1979 by a few Edo-Aire employees, who brought an interesting approach to the business. S-TEC utilizes the building-block concept, which allows you to start with the single-axis system and later add, item by item, the pitch axis, electric trim, altitude and vertical speed preselect, and flight director.

Their pricing structure allows you to upgrade without a large penalty in equipment cost. That is, you can buy a single-axis system, then if you decide later that you want the second axis, you can purchase a mod kit, and the total price will be not much greater than if you had bought a two-axis system to begin with—unless, of course, there has been a price hike in the intervening time. However, the piecemeal method will undoubtedly result in somewhat higher installation charges.

S-TEC also makes a pitch stabilization system that can be used with other makes of single-axis autopilots. The two systems do not interface—one flies roll and the other independently flies pitch—but together, they provide the capabilities of a two-axis autopilot and allow you to upgrade without scrapping your present system. If your roll system dies later on, you can get a mod kit that will upgrade the S-TEC pitch system to a full two-axis system. This will cost more, in total, than buying the two-axis system to begin with.

As I pointed out earlier, the S-TEC systems are rate based, using the turn coordinator for lateral control instead of the attitude gyro more commonly utilized. Thus, your confidence level will depend partly on the reliability index you give to electrical systems versus air-driven systems. If the DG fails, the S-TEC system will continue to function in every respect except for heading select. If the electrical system goes, you lose the entire autopilot.

The turn coordinator has a gyro whose spin axis is inclined 37° so the instrument can sense both bank rate and turn.

In addition to the standard turn rate, there are two soft modes initiated in sequence during capture of a VOR radial. The first capture mode reduces turn-rate capability to 45 percent of standard. After 1 min, the capability is further reduced to a maximum bank angle of approximately 2°–4°. This latter mode is eliminated when tracking the localizer. The glideslope can be intercepted from above if necessary.

The flight director has a two-cue display with a split-axis engage, enabling you to use only the roll portion if you wish.

S-TEC offers two stripped-down economy systems. The System 40 is a wing leveler with a radio tracker. Being a basic system, it does not have the automatic capabilities of a coupler; you must fly to within a few degrees of course on the desired VOR radial or a localizer, and then the autopilot will track it. The System 50 is similar to the System 40, with the addition of altitude hold. It does not have glideslope coupling. The 40 and 50 are single-box systems.

The System 60 is S-TEC's full-feature line (Fig. 13.1). The programmer and annunciator are combined in one box and the computer

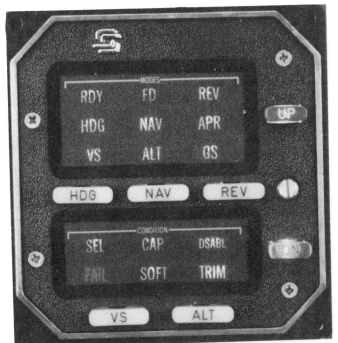

13.1. The System 60 annunciator/control panel. (*S-TEC Corp.*)

is remotely mounted. The ST60-1 is a single-axis autopilot and the ST60-2 is a two-axis unit. Both models have radial-intercept capability. They also feature a generous number of deviation warning signals. Flashing lights indicate low RPM in the turn coordinator, off-course needle displacement by 50 percent, invalid nav signal, off glideslope centerline by 50 percent, and the inability to maintain the selected climb rate.

Options include an HSI, a flight director, and an altitude/vertical speed selector.

Other manufacturers of lightplane autopilots are Brittain Industries, Inc., Astronautics Corporation of America, Inc., and Sperry.

Selecting an Autopilot. Of course, your choice will be limited to whatever equipment has been STCd for your plane; it is hoped you'll have several options.

Remember, you are choosing a copilot you will probably live with for a long time — so choose carefully. As we have discussed, there are all levels of sophistication available. The upper-end models will normally be of greatest interest if you do a lot of single-pilot IFR or busy terminal area flying — or if you simply like high-tech automation and are willing to pay for it.

One thing to bear in mind: an autopilot can be the most maintenance-prone item on your panel. Some are worse than others. It's a good idea to talk to avionics dealers and find out where the lemons are.

Try before you buy, if you can rent or get a ride in a plane that has the system you're interested in. And remember, most autopilots do a pretty good job in calm conditions. For a real test, pick a day with some turbulence. When a gust picks up a wing, notice whether the autopilot corrects smoothly. If it's a two-axis autopilot, make some climbs and descents. Observe how it levels out when you engage ALT (altitude); you can expect some overshoot, but it should settle back on the altitude you selected. And it should hold onto that altitude, even in some updrafts and downdrafts.

Try a VOR radial intercept and check for smoothness and overshoot. If the autopilot has a soft mode, allow it to take effect and then simulate scalloping by playing with the OBS. Also, note the behavior during station passage. Intercept the localizer at a good angle, and watch for overshoot. If it has pitch control, observe the way it intercepts and flies the glideslope.

Then do it again with another brand and compare performances. After all, if you were hiring a copilot, wouldn't you want to check him out pretty carefully?

A couple of thoughts about operating an autopilot. Carefully

AUTOPILOTS

read the owner's handbook *and* the supplement that is part of your airplane flight manual. Sometimes there are restrictions, such as flap limitations, that apply to specific aircraft models.

Preflight the autopilot before each flight, regardless of whether you intend to use it. Remember, it is connected to the controls of your plane, whether it is turned on or off.

If you encounter severe turbulence, it's a good idea to disconnect the altitude hold, and possibly the other modes as well.

Do not let the autopilot become the pilot in command. Monitor its actions. Even though it may be designed to warn you of its own system failures, gyros have been known to get tired and hand out false information. Cross-check from time to time with the various instruments on your panel, just as you would if you were operating the controls.

Check the gyro air filter regularly, especially if anyone smokes in your plane. Make sure the autopilot system is included in your 100-hr or annual inspection. The servos should be checked for capstan or cable wear, and for the proper tensioning of the primary control cables and servo bridle cables.

Earlier in this chapter, we touched on the HSI and flight director. Let's take a closer look at them now.

The HSI. *HSI* stands for Horizontal Situation Indicator (Fig. 13.2). To add semantic confusion, some manufacturers have their own names for the instrument. King, for example, calls it a Pictorial Navigation Indicator, or PNI, while Century opts for Navigation Situation Display, or NSD. I'm not sure why this is so; I'm just glad they don't also make ASIs, VSIs and the like, or we'd have to carry glossaries in the chart case.

By any name, it's a good instrument to have in your panel, with or without an autopilot. The HSI replaces the directional gyro and the course deviation indicator of the Number 1 Nav. It is not cheap—prices run over $5000 new and more than $3000 for a reconditioned unit—but it's an invaluable navigation aid, especially for instrument flying, and it makes the autopilot more automatic.

A major benefit of the HSI is that it gives you a clear picture of your plane's position in relation to its magnetic heading and the VOR radial or localizer it is tracking.

To see how the HSI helps make your autopilot flying easier, let's compare it with the DG it replaces. Suppose you are flying due north and want to intercept a 40° course to a VOR station. If you have an HSI, you turn the course pointer to the 40° setting on the compass card and go about your other chores. When the deviation bar centers and the autopilot captures, it will turn the plane to a heading of

13.2. A single-cue flight director (top) and HSI (bottom). (*King Radio Corp.*)

approximately 40°, because that is the error signal it was given via the HSI. Once the error has been corrected, the autopilot will keep the plane on track by responding to signals from the left-right deviation bar.

If you don't have an HSI, the autopilot will not know what course you want to take up. Therefore, upon intercepting the radial, you must move the heading bug on the DG to the course you want to fly. If you don't move the bug, the autopilot will still turn the plane in response to the CDI's left-right commands, but if the error between the DG bug and your course line exceeds the autopilot's built-in crosswind correction, it will track off to one side.

AUTOPILOTS

It helps to remember that an autopilot needs the same information you use when you're navigating. For example, when you are manually tracking a VOR radial, you turn to a specific course that will put you on that radial, and then you turn right or left to stay on the track. You need a reference course to begin with, and the autopilot needs that also. According to one autopilot manufacturer, many pilots don't take this into account and therefore don't fully understand how the autopilot operates.

There's another significant HSI benefit, called all-angle intercept. This means that with an HSI, you can use the autopilot to intercept a VOR radial at any angle you choose.

Here's a practical example: typically, in a terminal area, you are in a radar environment, talking to approach control, and will be given vectors to intercept your final approach course. With all-angle intercept, you can put the autopilot in the approach-armed mode and maintain the heading you've been given to intercept the final approach course. When the needle centers, the autopilot captures automatically.

The CDI/DG combination doesn't give you that flexibility. As soon as you arm the approach mode, the autopilot will make a 45° intercept to the final approach course. So to keep to your assigned heading, you have to remain in the heading mode.

Another example: you take off from an airport and want to intercept a VOR radial. The radial is close by, so the best procedure is to use a small intercept angle, which you can do with the HSI. But if you have only a CDI to work with, and you arm the autopilot to capture the radial, it's going to put you on that 45° intercept, which will be the long way around. Here again, the HSI lets you fly automatically the same way you would choose to fly manually.

The Flight Director. This is another one of those devices whose operation is often misunderstood.

The flight director is an aid to hand-flying the airplane with the same precision that the autopilot would have if *it* were doing the flying. In fact, the flight director presents to you, in visual cues, the same commands the autopilot issues to the servos when it is engaged. It utilizes a display called a steering horizon, which is basically an artificial horizon with command bars.

There are two types of display in general use. One is the double-cue system. This has a vertical and a horizontal needle with rectilinear movement, much like a course deviation indicator. The other is the single-cue system. It has two bars laid out like an inverted V and superimposed over a delta, or triangle (Fig. 13.2). The double-cue system is more precise, and therefore a bit harder to fly accurately—

which may be the reason why some pilots prefer the single-cue system.

Here's an example of how the flight director works: you're making an ILS approach, hand-flying the airplane. Somehow, you manage to get to the left of centerline and above the glideslope.

If you were using a CDI as your reference indicator, you'd see the vertical needle was to the right of center and the horizontal needle was below center, so you would correct to the right and pitch down. Depending on how far off you were, plus any crosswind component, you'd have to estimate the amount of corrective action it would take to center the needles without undershooting or overshooting. When the needles centered, you'd be on course, on glideslope.

But you're using the flight director instead, and let's say it has the double-cue display. You'll still see a vertical needle to the right of center and a horizontal needle below centerline. But you don't have to "guesstimate" the amount of correction to make—that is being computed for you. Just "fly to the needles," and they will center as soon as you have cranked in enough control input that will *result* in the plane getting back on track. In other words, when the needles are centered, you may not be on track just yet, but as long as you keep them centered, you'll get on track and you'll stay on track.

In the same way, the flight director will lead you to a new heading or VOR radial. Just keep the needles centered (or, in the case of a single-cue system, keep the V-bars over the top of the delta), and you'll get where you want to go.

Some flight director systems have a go-around mode. You activate it by pressing a button, usually located on the throttle. This causes the display to command a wings-level, pitch-up attitude that is appropriate for your airplane at climb power. In some cases, initiating this mode will disengage the autopilot if it has been engaged. This depends on the way your autopilot has been certified. In addition to go-arounds, the mode can be used for takeoffs or other situations involving wings-level flight at climb power.

Not everybody is thrilled with the flight director. Here are some blunt comments from John Nixon, Director of Engineering at Century:

> Most people use flight directors at a time when they have the least business using them. It's the very time when they really ought to be paying attention to what's going on outside the airplane, rather than giving themselves a voluntary case of tunnel vision. Studies conducted among the airlines showed that even the airline pilot's scan goes to zero when using the flight director.
>
> If we had a choice, we wouldn't even offer flight directors, but we're not about to tell customers what they don't want. Some people like to use the

flight director in the final moments of the instrument approach, because it puts them "in the loop" before they have to flare.

The average pilot really isn't nearly as good as he would like to think he is. Almost without exception, I haven't seen a single pilot who can do as good a job on the ILS as any of our autopilots can. Using the autopilot leaves the pilot free to be a cockpit manager instead of the guy responsible for keeping the wings level.

I asked Nixon if it would be fair to say that, if the pilot insists on flying the airplane, he will probably do a better job with the flight director than without it.

With one qualification: if he doesn't fall victim to tunnel vision. If he will continue to scan the other instruments, the flight director will perform a useful service for him. And if he's bound and determined to fly the airplane anyway, at least the flight director provides the benefit of doing some thinking for him and reducing a lot of variables down to just two—fly left or right, or fly up or down.

That's one man's opinion, but it carries some weight, because John Nixon is very much in the autopilot business. You may value the flight director more highly than he does, especially if you like to keep your hand in on precision approaches. The device may also pay for itself by helping you fly the plane if you get into turbulence so rough you have to disengage the autopilot.

But whenever you use it, you should always be aware of the potential danger of fixating on its display—or, for that matter, on any instrument or situation.

14. Mods

Spoilers, full-span flaps, and drooped ailerons...47 percent further on the same tankful of fuel...Converting a tri-gear to a taildragger...225 mph Bonanza...Why don't the manufacturers make the mods?

(Jim Larsen)

ARE YOU a candidate for an airframe modification? You are if there's something you want from your plane that the manufacturer didn't build into it and there's an independent shop that's found a way to supply that need for your particular model of airplane.

Some mod shops specialize in updating older models of a particular type of plane to the specs of the current model. Other shops go further to provide improvements the factory never got around to designing into the airplane.

Here are the kinds of benefits now being offered by various mod shops.

Performance Improvements. These run the gamut from shorter takeoff and landing distances and more precise control at slow speeds, to increases in rates of climb, cruise speeds, service ceilings, and, in the case of twins, single-engine performance. These benefits are wrought through the installation of gap seals, fairings, stall fences, better induction systems, leading edge cuffs, and new wingtips. Some of the more complex systems involve such exotica as spoilers and redesigned flaps. A speed improvement, of course, brings with it the important fringe benefits of added fuel economy, range, and/or payload.

If the mod shops can work their magic to create significant improvements over the factory specs, an obvious question arises: why don't the aircraft manufacturers make these changes on their own products? I asked this question of some of the mod shop owners, as well as executives of the leading plane makers, and the consensus follows.

The modifiers believe that the aircraft manufacturers have little incentive to make major design changes, unless they are being prodded by competition. As for adopting improvements that have been made by outsiders, there's the "NIH" (not invented here) syndrome at work. What's more, the smaller shops, with their relatively low overhead, can work their changes for considerably less cost than the plane makers themselves.

For their part, the manufacturers feel many of the mods alter the characteristics of the planes in ways that make them desirable only to a select part of the market. And some of the mods, they believe, either are not desirable at all or their benefits are overstated. In sum, the manufacturers take the position that if they felt the design changes would benefit most of their customers, they would make those changes.

The chief engineers of both Beech and Piper made the interesting point that a mod shop can get FAA approval to alter an airplane more easily than the company that designed and built that airplane. This doesn't seem reasonable, since the manufacturers have the foundation of research, development, and testing that got them the type certificate in the first place—but then, the FAA isn't always reasonable.

It's also fair to state that at least some of the manufacturers have made meaningful improvements over the years. An outstanding example is Roy Lopresti's transformation of the efficient but cluttered old Mooneys into the slick 201 and 231.

Utility and Cosmetic Improvements. Mods that add to a plane's ease of operation, comfort, and appearance can be as meaningful as those that improve performance. For example, the panels of older planes often reveal a hodge-podge of instruments whose placement borders on the whimsical. A panel update can organize the instrumentation, switches, and circuit breakers, while at the same time creating room for more radios and accessories. More effective panel lighting can be incorporated as well.

Also, some of the current aircraft have such amenities as third windows and one-piece windshields the predecessor models were not blessed with. There are shops that will provide a face-lift, making it hard to distinguish a plane that has been around for a few decades from one just off the assembly line. The picture can be completed with a new interior and extra soundproofing.

Some mods are available in kit form and can be installed by any competent repair facility, while others must be done on the premises of the modifier. If you buy a mod kit for installation in the field, ask the kit manufacturer to recommend a shop in your area with experience installing his products. Otherwise you might end up paying for the shop's learning curve.

Will a Mod Pay for Itself? Modifications can run from a few hundred dollars to many thousands. Flap and aileron seal kits will range in price from about $225 to $750, depending on the kit manufacturer and the airplane. A complete speed modification, including an assortment of fairings and gap seals, redesigned cowling, and induction systems, etc., can cost $11,000-$13,000 installed on a single-engine plane. Intercooler kits price out at $5000-$8000 uninstalled; a set of speed brakes is $3000 plus labor; and if you want to make a nice, old-fashioned taildragger out of your tri-gear Cessna, the parts will come to about $2000.

Can the expense be justified on a cost-effective basis? Probably not. You are buying greater capability, but you can expect to have a long wait before a speed mod, for example, will pay for itself in fuel savings. Let's take a couple of cases in point.

You can get a complete speed package for a Bonanza that will, according to the mod shop, increase the plane's 75 percent cruise from 200 to 220 mph, an improvement of 10 percent. The price of this package is $13,000. You could use this newfound efficiency in one of two ways.

YOU COULD SAVE FUEL. You could throttle back from 75 to 59 percent and still get 200 mph, while saving 3 gph. Figuring fuel cost at $1.90/gal, that's a saving of $5.70/hr. Let's say you put 400 hr a year on your plane. At that rate, you'll save $2280 annually. But you've lost the use of that $13,000, which could be drawing, say, 10 percent in a certificate of deposit account. So subtract $1300 from $2280, and you have a net gain of $980. At $980 per year, it will take a little over 13 years to recover your $13,000. If you fly 200 hr a year, the fuel savings will not compensate for the loss of interest on the investment.

MORE LIKELY, YOU'LL WANT THE SPEED. You're not really spending all that money to throttle back and save fuel, except possibly in some marginal situation when you are trying to squeeze some extra range out of the plane. Most of the time you'll be taking advantage of the higher speed, so let's see how cost effective that is. For simplicity's sake, we'll assume you'll be covering the same distance in 10 percent less time. If your direct operating cost is $55/hr (fuel, oil, engine, and prop maintenance/overhaul), you might say you're saving $5.50/hr by shutting down 10 percent sooner. Therefore, the dollars and cents come out to about the same as in our previous example.

Would a simpler, less expensive mod package be more cost effective? Let's see. You can get a flap and aileron gap seal mod that will provide about a 3 percent speed increase on an Archer for approximately $1200 installed. Figuring a direct operating cost of $30/hr, the mod will save you about $.90/hr. At 400 hr per year, that comes to $360, minus $120 interest you've lost on the $1200 investment. So your net gain is $240, and the mod will thus pay for itself in five years. At 200 flight hours per year, it will take 20 years for the mod to break even.

A powerplant mod is usually accomplished by swapping the factory engine for one that adds horsepower and/or turbocharging, and this change is almost certainly going to add to operating expenses and be less cost effective than the original equipment.

WHAT ABOUT INCREASED RESALE VALUE? It's hard to say how much of your mod investment will be returned when you decide to dispose of the bird. On the one hand, whenever you sell something that's a bit of an oddity, you inevitably find yourself appealing to a relatively select group of buyers. A lot of prospects may not share your enthusiasm for, and confidence in, a plane that has been tweaked and altered by persons who were not responsible for the aircraft's design and manufacture.

Conversely, if your mod has lived up to expectations, you will be able to offer a plane that is superior to its factory siblings, and you might be able to sell it at a premium.

But to avoid overly high expectations, consider your mod to be an expenditure for the sake of performance, plus whatever satisfaction you might get out of having a plane different from nearly everyone else's.

Choosing Mods and Modifiers. Modifiers run the gamut from kit manufacturers who don't do any installation work, to repair shops that do mods as a sideline, to facilities whose main business is the design and installation of aircraft modifications. The majority of mods require STCs (supplemental type certificates), often involving extensive testing and paperwork for each model of airplane for which the mod is intended. Other modifications, which are not considered by the FAA to alter the performance characteristics of the airplane, can be installed by the simpler Form 337 procedure.

If you are ordering a kit, it's important to know what kind of paperwork is needed to satisfy the FAA General Aviation District Office in your area. If the modifier has obtained an STC, is it a multiple-installation STC or a one-time STC? If it's the former, it should be acceptable to your local GADO with no problems. If it's the latter, your installer will have to get an STC for your particular plane, which could run up the cost.

I've met a number of modifiers over the years, and they have all appeared to be dedicated people who truly believe in what they are doing, and whose products and workmanship are of good quality. However, some of them, in their zeal, tend to overestimate the performance benefits their mods provide. This may be due in part to the modifier's lack of the extensive and sophisticated testing procedures that the airframe manufacturers use in generating their specifications.

Before going to the expense of a mod, I suggest that you ask the modifier for a list of customers; then, reach out and touch a few of them. Most pilots are happy to talk about their planes, but remember, some of *them* tend to exaggerate, too. So try to find out whether their testimonials are based on guesswork or careful evaluation.

The Modifiers. Here is a sampling, in alphabetical order, of companies that offer modifications. Please note that the information, including performance claims, is supplied by the modifiers, and I cannot attest to their accuracy.

ADVANCED AIRCRAFT CORPORATION. They have a turboprop modification for the Cessna P210, using a Pratt & Whitney PT6A-135 engine. Specifications include a 2050-fpm rate of climb at 4000 lb, a cruise of 262 kn at 23,000 ft, and a range of 1008 NM with 140 gal of fuel (including 52-gal aux tanks.) Advanced will sell a complete airplane, with redesigned interior, or they will make the power conversion on the customer's aircraft. The company is still in the process of testing and expects certification shortly.

AERO-TRIM, INC. Aero-Trim offers an aileron trim system, which unfortunately is not found on most single-engine aircraft. Even if a plane is properly rigged, it will be wing-heavy during most of its flying time, because there is almost always more fuel in one wing tank than the other. Correcting this condition with rudder causes an inefficient sidewise motion to develop. Aero-Trim's system utilizes an aileron trim tab and an electrically operated servo to solve that problem. The system is STCd for well over 200 types of aircraft. A rudder trim system is also available for the straight-tail Bonanzas.

THE AMERICAN NAVION SOCIETY. Navion provides a number of mods for Navion owners. The performance mods include speed fairings, raised gear doors, aileron-balance kits, fiberglass nosewheel fairings, and a fuel-filler door cover. Numerous engine and accessory kits are also available. Among the utility and cosmetic mods are bubble windshields and panoramic one-piece side windows, headliner kits, velour upholstery, instrument-panel kits, fresh air systems, and control wheels. They also supply Brittain tip tanks, which hold 20 gal of fuel per side.

AMEROMOD CORPORATION. Ameromod offers engine conversions for the Grumman-American two-place models, with power choices of 125, 150, 160, and 180 hp. They will also turn a Cheetah into a Tiger. In addition, Ameromod installs 10- and 20-gal aux fuel systems for the G/A two-placers. Also, they will install a Sensenich prop for the Tiger that eliminates the yellow arc between 1850 and 2250 RPM and dispenses with the AD that requires periodic hub inspection.

The company has two mods that turn the Cessna 152 into what Ameromod calls a Sparrowhawk. The first is a prop and spinner replacement that allows the engine to develop 115 hp at 2700 RPM for improved takeoff performance. The second is a conversion of the existing engine with a piston change and a baffle system mod, producing 125 hp and resulting in cruise speeds of 135 mph at 75 percent power.

The Piper Tomahawk is given this same mod, along with a dorsal fin and wheel fairings. Future Ameromod plans call for a turbocharged 180-hp Tiger, plus a 200-hp fuel injected conversion with constant speed prop for all four-place G/A models.

AVCON AND BUSH. They are two related companies that install a 150-hp Lycoming in the Cessna 150/152 and a 180-hp Lycoming in the Cherokee 140, 150, 160, and Warrior, as well as the Cessna 170, 172, 175, and early 177.

AVIATION ENTERPRISES. This company has conversions for Bonanzas, Barons, Travel Airs, and Dukes. Their performance mods include flap and gap seals, a speed-sloped windshield, and a fiberglass tailcone that can house the marker beacon antenna or ADF loop for added streamlining. For customers interested in utility and cosmetic mods, they install instrument panels, sliding seats, windshields, and third windows. Their instrument-panel update is similar to the factory V35B panel and includes new engine and fuel gauges. One noteworthy feature is that the flap handle has been placed considerably higher on the panel than the gear handle, to help reduce the possibility of inadvertent gear retraction on rollout.

A very popular product is their chrome-steel towbar that fits into the nosegear axle instead of clipping onto the tow pins, presumably making the entire operation less of a contest.

BERYL D'SHANNON. D'Shannon specializes in making older Bonanzas look and perform like the newer models. They install speed-sloped windows, extra side windows, tip tanks, instrument panel updates, and seating mods. They also put on the third spar mod for V-tail Bonanzas. D'Shannon will replace the old front bench seat with sliding seats and will put a couch in the back on request. Their performance mods include gap seals and tip tanks with winglets, as well as IO-470 and IO-520 engine installations. They are planning to get an STC for the 300 hp engine Beech is now installing on the A36. Mods are also available for the Comanche, Swift, Mooney, and Cessna 310, 320, and 337.

BOB FIELDS AEROCESSORIES. Fields provides an effective solution to cabin noise: an inflatable door seal. Three systems are available. The deluxe electric is fully automatic in flight with an electric power-supply module. The deluxe manual is operated by a hand-held bulb, similar to that used for blood-pressure tests. An air-flow control valve is mounted on the panel. The economy model is similar to the deluxe manual, except the air-flow control valve is on the bulb itself. The seal is STCd for most Cessna singles, the 310/320, Bonanzas, Barons, Mooneys, and a number of Piper models.

CAMERON AIRCRAFT INTERIORS. Cameron does upholstery and side panels. In addition, they have a full-time cabinet man who works mostly on executive aircraft, but for smaller planes will install a Jepp rack or a service center that will hold a thermos bottle, cups, and ice. These items are custom made to the aircraft owner's specifications.

COLEMILL ENTERPRISES. They turn Navajos into high-performance Panther Navajos, with 350 hp Lycoming TIO-540 engines and four-blade Q-tip props. They also install 300 hp engines on the Cessna 310 and Beech B55 Baron.

CS INDUSTRIES. CS turbocharges Cessna 180s (Models F through J), 182s (H–Q) and 185s (E and F). The company uses AiResearch systems, including automatic wastegate controllers, and the engines are converted from carbureted to fuel injected.

CUSTOM AIRCRAFT CONVERSIONS. They make kits that turn a 150 or 152 into a "Texas Taildragger" 150. The benefits of the tailwheel conversion include a 10- to 15-mph increase in cruise due to drag and weight reduction. Also, the rate of climb will improve by about 65 fpm, and, of course, prop clearance will be better. Maintenance costs should go down, because owners will no longer have to cope with the infamous nosegear shimmy. Installation takes 45–55 hr. A taildragger conversion for the 172 is available as well. Also, Custom has a 180-hp engine conversion for the 150/152 and is working on a 150-hp changeover. Another 150/152 mod is a pair of 40-gal long range fuel tanks, which are swapped for the existing 26-gal tanks.

FLIGHT BONUS. Flight Bonus specializes in fixed-gear Skylanes, offering a number of drag-reducing kits (Fig. 14.1). Depending on the year of manufacture, a Skylane could be modified with up to five kits. These include a streamlined nose-gear fairing, streamlined main gear, wing strut fairings, cowling and prop-spinner closures, and flap well and aileron gap seals (also available for the 172). Skylanes made from 1967 to 1971 can use all five kits, for a total cruise-speed improvement of 20 mph. If the plane owner prefers fuel economy to increased

14.1. A Skylane with specially designed strut and wheel fairings. (*Flight Bonus, Inc.*)

speed, he can reduce power to 50 percent and achieve the same cruise speed he had been getting at 75 percent power.

Skylanes of other vintages, from 1956 to 1980, can use from one to four kits. Installation of all five kits takes about 80 hr. Flight Bonus does not perform installations, but the work can be done at a repair facility on their field at Arlington, Texas, and at other shops throughout the country.

FLINT AERO. They are in the business of building aux fuel tank kits for Cessnas. They have internal wing tanks of 23 gal usable capacity for 150s through 185s, plus early 210s. External tanks replace the wing tips of early 206s, the 207, and the 210 D through F; adding 27-gal usable capacity and 30–36 in. of wingspan. Later-model 206s take a slightly larger tank. Fuel tanks for the cantilever 210s add 32.5 usable gal. Flint does not do installations, but the work can be done at a shop at nearby Gillespie Field in San Diego. Flint estimates that installation time should be 35–40 hr.

HORTON STOL CRAFT. Horton installs wing airfoil modifications, conical wing tips, stall fences, and aileron gap seals on Cessna singles and the Cessna 336 and 337 Skymaster. They also produce a kit for Cherokee singles with the older Hershey Bar wing. The kit consists of wing-foil modification, dorsal fin and fairing tip, stall fences, vortex generators, and drooped wing tips. The purpose of these mods is to improve takeoff, slow-flight, and landing characteristics.

ISHAM AIRCRAFT. They sell STOL kits for Cherokees with Hershey Bar wings. The kits consist of wing extension, wing tips, a dorsal fin, and, for the fixed-gear models, a stabilator extension. These kits are said to enable the modified Cherokee to climb at a better rate and cruise faster than the equivalent Piper with the Warrior semi-tapered wing. Another benefit is improved aileron response at slow speeds.

In addition, Isham markets a kit that increases the horsepower of the Cessna Hawk XP from 195 hp to 210 hp. This is accomplished by exchanging the governor, tach, and mp/fuel-flow gauge, plus installing a new fuel-flow placard and resetting the propeller low-pitch stop adjustment. Benefits include a shorter takeoff roll and increased rate of climb.

KNOTS 2 U, INC. They provide flap seals and aileron seals for Cherokees with either Hershey Bar or semitapered wings, claiming owners report speed increases of 6-10 mph. Their T-tail Lance mod also includes stabilator seals and allows the plane to be rotated at 5 kn less than the unmodified aircraft. A comprehensive package of seals, fairings, and other cleanups is offered for the Comanches, resulting in reported cruise speed increases of 6-15 mph and improved handling characteristics.

KNOTS 2 U has also developed a stainless-steel stabilizer plate that reinforces the leading edge of the stabilizer on V-tail Bonanzas, Models C35 through V35B.

LAKE AERO STYLING AND REPAIR. Lake is in the business of updating older Mooneys to give them the higher-performance benefits of the slicker 201. They install such mods as flap gap seals, aileron gap seals, and dorsal-fin vertical seals; fairings for the wing tips, wing roots, and aileron push rods; a redesigned cowling, and just about every other device designer Roy Lopresti came up with to make the Mooney go faster on the same horsepower. In addition, Lake is modernizing the instrument panel of older Mooneys, regrouping instruments into the basic T, and placing the VOR indicators in the pilot's line of sight. Lake is also seeking an STC for the single-piece smooth belly incorporated on the 1984 Mooneys.

LAMINAR FLOW SYSTEMS. They provide mod kits for the Seneca, consisting of fairings that cover the screws and rivets on the wings, flap gap seals, flap track fairings, wheel well fairings, and resculpturing of the inboard leading edge. STCs have also been obtained for the wing fairings for PA 28 and PA 32 models. A modified Arrow has shown a gain in rate of climb of over 180 fpm and airspeed increases ranging between 13 and 18 mph.

LOMPOC AERO SPECIALTIES. Lompoc sells custom inner-window kits for Cessna swing-out windows. These kits double the thickness of the pilot and copilot windows and create a dead-air space between the panes. According to a sound-level test conducted in a 182Q, this resulted in an average noise-level drop of approximately 4.9 db, a substantial reduction. The insulation also adds to the comfort of the occupants by allowing for a greater level of warmth in the upper part of the cabin under cold weather conditions.

MACHEN INDUSTRIES. Machen installs 325-hp and 350-hp engines in the various Aerostar models. The 325-hp installation is a reworked version of the standard 290-hp powerplant. Benefits include improved climb and cruise performance along with an increased pressurization differential in the pressurized aircraft. A baggage compartment auxiliary tank is available. The turbocharged 350-hp engine is also installed in Bonanzas.

MARSH AVIATION. They put the Allison Model 250B 17C turboprop powerplant into the Cessna 206, 207, and 210 aircraft.

MET-CO-AIRE. They sell Hoerner wing tips and tiptanks. The Hoerner contour design effectively increases the wing area and the aspect ratio and reduces parasite and induced drag. Performance improvements of the wing tips include a 60-fpm increased rate of climb, a 3- to 5-mph cruise-speed increase, a 4- to 5-mph reduction in stall speed, and a greater overall stability. The wing tips are available for Cessna singles, Model 35 Bonanzas, Cherokees, Comanches, Aztecs, and Apaches. Installation time is 2 hr.

The tip tanks are made for the Aztecs and Apaches and contain 48 gal of fuel. Installation time is approximately 25 hr.

MILLER AIR SPORTS. Miller updates older Mooneys with the 201-style windshield, fiberglass cowling with improved baffle system, and large spinner.

J. W. MILLER AVIATION. They replace the 160-hp engines in Twin Comanches with 200-hp Lycoming IO-360 powerplants, in either normally aspirated or turbocharged versions. Other mods include a lengthened nose, auxiliary tanks, dual brakes, and a one-piece windshield.

PRECISE FLIGHT. They make spoilers, not for roll control, but for descent control. The purpose of the spoiler system is to allow a descent path much steeper than normal without picking up excessive airspeed. This enables the pilot to make a steep approach over high obstructions, using takeoff flaps for a safer go-around. He can also achieve a more precise touchdown point and reduce tire and brake wear by dumping lift upon ground contact.

Other advantages include the ability to stay in smooth air longer before descending to pattern altitude, rapid descents without overcooling the engine, staying above the vortex turbulence of heavy aircraft while making approaches, and the ability to make normal landings in the event of flap-system failure.

Speed brake systems are available for the Cessna 180, 182, 185, and the 200 Series, as well as the Model 33 and 35 Bonanzas, Turbo Arrow, Turbo Dakota, Seneca, Mooney Models 20, 201, and 231.

The systems operate on engine vacuum, using their own auxiliary vacuum arrangement. This has been designed to serve as a backup source that will operate the aircraft's vacuum instruments in the event of a vacuum-pump failure.

RAM AIRCRAFT MODIFICATIONS. RAM installs the 100-octane Lycoming 0-320 engines in 1968 to 1980 Skyhawks, the Cherokee 140s, and PA 28-151 Warriors. The company also does engine updates on 200, 300, and 400 Series Cessnas.

RAMSHEAD EXCLUSIVES. The company offers custom-made sheepskin seat covers for all aircraft models. The covers are made of 1 in. thick Australian sheepskin and are available in 14 colors.

R/STOL, INC. This company produces all the mods that had been offered by Robertson, including drooped ailerons, contoured-wing leading edges, stall fences, sealed ailerons, and automatic pitch compensators. These conversions are available for Cessna singles and Senecas, Aztecs, and Twin Comanches.

The drooped-aileron mod is a flap-aileron interconnect that causes the ailerons to deflect downward when the flaps are lowered. This gives a full-span flap effect, which generates extra lift and enables the pilot to maintain less of a nose-high attitude—for better visibility and improved engine cooling. Also, stall speed is lowered.

Another R/STOL mod does away with ailerons completely, to give the airplane the lift advantage of full-span flaps.

Roll control is provided by spoilers (Fig. 14.2). This mod is performed on V35 and A36 Bonanzas and on the Seneca I and II. The full-span flaps reduce the liftoff speed of the Bonanzas from 80 kn to 57 kn. Best angle-of-climb speed is reduced from 83 kn to 62 kn, and the 50-ft barrier is cleared in 1350 ft instead of 1800 ft. Best rate of climb is achieved at 78 kn, down from 96 kn. The same type of mod on the Seneca I and II offers comparable takeoff and climb improvements, as well as reduced accelerate-stop distances.

R/STOL installs Fowler flaps on the Cessna 310, 340, 401/2, 414, and 421 (Fig. 14.3). The drooped aileron mod is also available on the 421C.

SEGUIN AVIATION. Seguin takes the somewhat lackluster Apache and rebuilds it from scratch into what they call the Geronimo. The airframe is stripped to its chrome-moly frame, which is deblasted, magnafluxed, reconditioned, and corrosion-proofed. The nose is ex-

14.2. In this mod, a spoiler replaces the aileron and permits the use of a full-span flap. (*Jim Larsen*)

14.3. Fowler flaps on a Cessna 310. (*Jim Larsen*)

tended either 16 or 31 in., providing a baggage compartment for better weight distribution and handling. Other improvements include Hoerner tips, fully enclosed wheel well doors, flap gap seals, engine nacelle and wing root fairings, a dorsal fin, and a larger tail.

Interior modifications include a ¼ in. thick wraparound windshield, for better visibility and noise reduction, and a custom interior with orthopedic seats and three-density soundproofing. A modern instrument panel has a double radio stack, an improved regrouping of instruments, and small gyros. An extended rear baggage compartment is built in, along with the forward compartment that comes with the nose extension.

For those customers who don't have their own Apache to modify, Seguin will supply one.

SMITH SPEED CONVERSIONS. They are dedicated mostly to Bonanzas, but they also work on some of the Beech light twins. Smith offers a flap and aileron gap seal package they say will double the roll rate, increase the rate of climb, and provide a 3- to 4-mph increase in speed. This is available on the Beech 33, 35, and 36 Bonanzas, the 55 and 56 Baron, and the Travel Air.

Another mod available for the Travel Air and 55 Baron consists of rigging and sealing the wings and gear doors, with benefits that include an increased roll rate, cruise-speed range, and rate of climb.

Smith has a complete aerodynamic drag reduction program for the Bonanza, consisting of flap and aileron gap seals, fairings, rerigging of the wings and gear doors, a redesigned cowling with ram air door, and removal or relocation of drag-producing elements. This provides a cruise speed increase of up to 20 mph.

As a safety item, Smith offers a stub spar stabilizer beef-up kit for the V-tail Bonanzas. This increases the strength of the stabilizer by 30 percent and relieves the torsional load on the spar. The stub spar is available as a kit. Installation by Smith includes checking the balance of the ruddervators, inspection of attach fittings and push-pull rods in the tail, plus examination of all the control cables for corrosion, chafing, or broken strands.

Smith is applying for an STC to install the 300-hp engine on Model 33, 35, and 36 Bonanzas. They are also planning a ram air intercooler for the turbocharged Bonanza.

SOLOY CONVERSIONS. Soloy provides turbine installations for Cessna 185s, 206s, and 337s. Detroit Diesel Allison turboshaft engines in 420-shp, 500-shp, and 650-shp configurations are utilized. The propeller and reduction gearbox are especially designed to reduce vibration.

SUPERPLANE, INC. Superplane has developed speed fairings, flap seals, and gap seals for the Cessna 206 and plans to offer these mods for the 172 and 182 in the future. The normally aspirated 206 will show a 15-kn improvement at 6500 ft if rigged properly. The Turbo 206 will cruise at 177 kn at 12,000 ft.

TURBOPLUS. They have designed intercooler systems to prevent the lost efficiency and high maintenance costs resulting from the overheating of turbocharged engines. In addition, the engine's critical altitude is increased as much as 6000 ft, and fuel consumption is significantly reduced. Intercooler systems are available for the Seneca II and III, the Turbo Arrow, Turbo Dakota, and Mooney 231.

Turboplus also offers pressurized magneto kits for Senecas and Turbo Arrows and is a distributor for the Precise Flight speed brakes.

UNIVAIR. They have a taildragger conversion for those who wish to make Pacers out of Tri-Pacers. The Colt can be converted as well. A Lycoming IO-360 engine conversion with Hartzell constant-speed propeller is available for the 108 Stinson series, and the Continental 0-200 is offered for the Reed Clipped Wing J-3 Cubs with metal-spar wings.

WINGS WITH SPRINGS. They provide a Cessna wing strut modified to include an air/hydraulic shock absorber. Its purpose is to reduce gust loads, provide greater stability and safety in rough air, and, because of higher dihedral, offer a higher resistance to stalls. STCs have been issued for Cessna 170Bs, 172s, 175s, 180s, and 182s up to 1972.

There's More. The above listing is by no means complete, but it should give you a pretty fair idea of the things that can be done to make an airplane do more of what you want it to do. Information on other modifiers can be found in aviation publications. *Plane & Pilot* magazine has a special mods issue each year, usually in June or July, and *AOPA Pilot* does occasional mod roundups. Also, a lot of modifiers advertise in *Trade-a-Plane.*

15. Airplanes of a Different Sort

The composite planes... The penalties of being different... Burt Rutan damns the twins... Bill Lear's final dream.

IN THIS CHAPTER, we'll take a look at a few airplanes that are unique and possibly, in this surprisingly conservative industry, ahead of their time. A couple of them have reached certification and even limited production. All of them are trying to clear a formidable barrier of costs associated with the R & D, certification, production, and marketing of an airplane in today's complex society.

These planes have been shaped by creative ideas. Even if the planes don't survive... I suspect the ideas will.

The Windecker Eagle. At this writing, the Eagle is the only all-composite single-engine airplane that has been certified by the FAA (Fig. 15.1). It was conceived by a dentist named Leo J. Windecker and certified in 1969. Only five Eagles were manufactured, after which Windecker Industries shut down due to financial difficulties. However, Jerry Dietrick, a businessman who had one of the Eagles and admired it greatly, took over the assets of the company in 1977 and has been trying ever since to raise money to get the plane back into production.

The composite used in the Eagle consists of several plies of nonwoven unidirectional fiber, with foam and laminate stringers for stiffening, and epoxy bonding. The fuselage material is laid up in two molds — one for the upper half and one for the lower half — and the two sections are bonded together. The upper- and lower-wing skins are similarly molded and bonded to the wing spars. The plane has two doors.

The Eagle as it is presently certified is similar in size and performance to the smaller Bonanza. Both planes have a gross weight of 3400 lb, and the cruise and stall speeds are within a few knots of each other. They also use the same Continental 285-hp powerplant. When it was being produced in 1970, the Eagle listed for about five percent more

15.1. The Windecker Eagle composite airplane. (*Composite Aircraft Corp.*)

than the Bonanza—$41,500 versus $39,250. (Those were the good old days; today, you could spend $80,000 for the *optional equipment* in a new Bonanza.)

Composite construction offers a number of significant advantages over the traditional aluminum construction. The composites have greater strength-to-weight ratio. They form a smooth surface, without the drag-producing rivets, lapped seams, and other deformations present in most metal aircraft. They can be molded in complex shapes, giving the designer more flexibility. Composites are not subject to corrosion and do not need to be painted. Also, antennas can be imbedded in the material, further reducing drag—although there is some debate over whether they are as effective as the conventional projecting antennas.

According to a paper written by George Alther, Chief Engineer of Dietrick's Composite Aircraft Corporation, the energy requirements for fabricating a composite airplane such as the Eagle are only 32 percent of the amount used in making a comparable aluminum plane. There are potential labor savings as well, due to fewer parts and assembly operations, but no real data is yet available, since only a few composite aircraft have been produced.

There is also a lack of information on the important questions of strength retention and bonding integrity over a long service period. However, Alther cites the following incidents involving Eagles:

1. Examination of a prototype aircraft which crashed during spin testing revealed no bond failures, either composite-to composite or metal-to-composite. The entire structure remained remarkably intact.
2. An extremely hard landing, where a panel-mounted G-meter registered 11 Gs, produced no damage to the aircraft structure other than to metal landing-gear components.
3. Three gear-up landings have occurred. In each case, the aircraft was repaired and returned to flying condition in a short time.
4. Severe in-flight turbulence has been encountered numerous times without structural damage. In one case, an installed G-meter recorded 5 Gs.
5. A fuel-cell leak was repaired on one aircraft, and it is suspected that this leak had existed for several years. No deterioration or contamination from the fuel was found in the wing's composite structure or bonds.
6. Examination of an aircraft, totally destroyed on the ground by a tornado, revealed no failure of any bonds.

Alther goes on to state that the five Eagles accumulated 8000 flying hours over a ten-year period, with no evidence of material or bond deterioration.

With all that composites have going for them, why are the major airframe manufacturers still making aluminum planes? First of all, some of the companies are now fabricating certain airframe parts, such as wing tips and cowl sections, out of plastic, and bonding is also being used on some wing and fuselage areas. Then there are the mostly composite corporate aircraft, such as the Starship and Lear Fan, that may be produced in the near future.

But for the most part, today's airframes are still constructed of aluminum sheets, riveted onto ribs and stringers, just like they were 40 years ago. Some years ago, Piper did design a composite plane called the Papoose, but it was never produced. I think the main reason for the manufacturers' resistance to change is the heavy investment in dies and jigs and machinery and know-how they have in aluminum fabrication. They are reluctant to discard all this and at the same time go through a burdensome period of retraining their labor force. This is why a lot of the impetus for composite airplanes has come from entrepreneurs who are outside the establishment.

The problem with this, as people like Jerry Dietrick, August Bellanca, and many others can tell you, is that newcomers have great difficulty in obtaining the huge amounts of capital required to put an airplane into production—especially if it does not follow tried-and-true methodology.

After years of trying to get financing to produce Eagle I, a somewhat improved version of the existing Eagle, Jerry Dietrick now intends to bypass that model and go into production on Eagle II, which was to have been the step-up model.

The Eagle II is an unpressurized turboprop plane with the same fuselage and four-seat capacity of the Eagle I. Dietrick feels that an unpressurized propjet with a price tag of $300,000 has a much better chance of selling than a pressurized single-engine propjet, such as the Beech Lightning or Smith Propjet, at about twice the cost.

Why no Turbo Eagle I? When I asked Dietrick that question, his answer seemed very subjective, but then, that's a characteristic of many entrepreneurs. It turns out he had an unsatisfactory experience with a turbocharged 180-hp Mooney. This was a retrofit, and the engine was not designed for turbocharging, as are the powerplants on the Mooney 231 and virtually all other factory-turbocharged planes. Nevertheless, Jerry feels that a turbocharged engine works too hard, is too costly to maintain, and produces too little benefit for consideration on the Eagle.

As for the status of Eagle II, I quote from a recent letter from Dietrick:

At last we can see the light at the end of the tunnel. In fact, it appears to be a spotlight. The time for this project is now.
Our Eagle/Allison turboprop conversion is ready to fly. We recognize a unique situation, in that the industry refuses to provide the business pilot with the aircraft he needs. The Eagle/Allison turboprop is in the process of supplemental type certification. Our business plan for production of 450 aircraft during the next five years is ready to be put in operation.

By the time you read these words, maybe we'll be seeing a single-engine propjet of composite construction, with a useful load of 2150 lb and a cruise speed of 220 kn. Then again, maybe not.

The Skyrocket. This sleek composite plane lives up to its name. It is very fast, cruising at 258 kn at 24,000 ft. It is powered by a 435-hp turbocharged and geared Continental and will carry six occupants in its pressurized cabin. The prototype is unpressurized.

The Skyrocket was designed by August Bellanca. He is the son of Giuseppe Bellanca, who created the fabric-covered, wood-wing Junior, Crusair, and Cruisemaster aircraft that later evolved into the Viking series.

As with the Eagle, the Skyrocket's fuselage is formed in an upper- and lower-mold, and the two halves are bonded together. However, the materials are somewhat different. The Skyrocket uses fiberglass with a core of honeycomb aluminum; the core on the production model may be Nomex.

The wing has a NACA 63_2-215 laminar-flow airfoil, and Bellanca says his is the first airplane to achieve laminar flow on 50 percent of the wing chord. A NASA test program on the airplane showed a drag coefficient of .016, which is more than half the typical general-aviation figure of .03. This contributes to good fuel efficiency.

In its present form, the Skyrocket has a gross weight of 3775 lb and a usable fuel capacity of 176 gal. According to Bellanca, the plane will carry three occupants and 50 lb of baggage approximately 2000 mi.

He also states that the pressurized airplane will not be much heavier than the prototype, because some new materials will be used, as well as weight-cutting manufacturing refinements. The production model will be equipped with dive brakes.

Bellanca hopes to make three versions. The piston-engine single will be priced at about $250,000. Then there will be a twin, powered by 380-hp Lycoming turbocharged direct drive engines. This is expected to cruise at about 300 kn and sell for $325,000 to $350,000. Next, a 315-kn turboprop single will go for about $500,000. All three planes will use the same basic six-place fuselage.

Now we come to the question of finances. Like other brash newcomers, the Skyrocket has had a turbulent fiscal flight. The plane began life in the 1960s on Long Island, then moved to West Virginia, where a group of investors was located. The prototype first flew in 1975, but there were the usual fund-raising problems. Bellanca feels the financial proposal did not have enough documentation as to certification and production costs. There was also some public skepticism about the Skyrocket's performance capabilities.

Bellanca broke up with the West Virginia group and purchased their stock. He has prepared an extensive business plan and is working with an investment banking firm. The Skyrocket has been moved to Delaware.

The certification programs for all three versions will be handled by an aerospace company, at a cost of about $3.5 million. Bellanca estimates a total outlay of $20–$25 million to get the planes into profitable production.

The Defiant. This was to have been the first Burt Rutan design to be produced on an assembly line. Rutan and the late Howard "Pug" Piper tried to interest aircraft manufacturers and others in producing the unique twin, without success. So the Defiant has gone the route of other Rutan designs, into the homebuilt market.

Although this book is not about homebuilts, I am including the Defiant in this chapter because it is a very interesting plane, and I share Rutan's feeling that it should be available as a production aircraft. Perhaps someday it will.

As is typical of Rutan's designs, the Defiant is a canard, with small wings up front and large swept-back wings, with winglets, at the aft end—much in the style of the VariEze and Long-EZ.

One major difference: it's a twin, utilizing the centerline thrust concept, with one engine pulling from the front and the other pushing at the rear. In this respect, it is, of course, reminiscent of the Cessna Skymaster. However, according to Rutan, there the comparison ends. When the Defiant first flew, in 1978, I chatted with Rutan, and this is what he had to say about the Cessna product:

> The Skymaster is a compromise airplane. It solved one of the problems and made some of the others worse. The ability to recognize which engine fails in the Skymaster is worse than the competition. You can't play "dead foot, dead engine."
>
> Its flying qualities at minimum control speed are obviously better than the competition. Look at the accident statistics and you'll find that the Skymaster is indeed better than the other twins, but not as good as the singles, because it's a worse airplane in the other respects. Its single-engine rate of climb, particularly with the back engine failed, is miserable. It has a lot of accidents just because pilots don't know the back engine has quit. So the Skymaster is a safer airplane than other twins, but not as good as it should be.

Rutan's original plan was to make a twin that could climb at gross weight with full back stick, gear down, and one prop windmilling. Here's how, in Rutan's words:

> To do this, we're looking at a big improvement in induced drag. There's no such thing as a real big improvement in parasite drag, because the current airplanes are reasonably good—the better ones, anyway. But for induced drag, there is some improvement available.
>
> Induced drag, which is drag due to lift, is negligible at maximum speed, but it represents half the drag at climb speed and most of the drag when you're below climb speed.
>
> So if we can lower the induced drag, we have an airplane that can get below best climb speed and not drop out of the sky. The VariEze is an excellent example of that. The best rate of climb is up there at about 95 mph. But the minimum power required is way back at 75 or so. At very low speeds the airplane requires very little power. You take a conventional Cessna, Cherokee, Seneca, or whatever, and if you get down to speeds approaching the stall, you have to come up quite a bit on the power to hold level flight.
>
> The Defiant is a more efficient airplane—more compact, with less induced and parasite drag. The current light twins are heavy, due to several reasons. One, they're not the most efficient, structurally. You go torsionally in the outer wing and cantilever the engine out, and then you have to take those

loads back into the fuselage, with all the torsion and all the twisting. Then you have a big cone on the back of the fuselage that does nothing but hold the tail, which includes a rudder that has to be enormous to be able to hold the minimum control speed.

My approach is to get the empty weight down. We don't do this by making the airplane small inside. Once you get behind the cabin, you're looking at a firewall and not a big long tube holding the tail on. The bending moments in the fuselage are much less than they are in the competition. Instead of having a big wing with the engines on them, and the tail pushing down, trying to bend the fuselage, the load and lift is distributed along the fuselage. So the weight of the fuselage goes way down.

Also, using composites saves us some weight. Another big weight-saving factor is that the Defiant doesn't have flaps. When you put flaps in, you have to have actuators, transducers, flap tracks, and you've got to strengthen the ribs. You put a piece of weight here, you have to chase that load all the way out to the fuselage, and it ends up adding a lot of weight.

This plane is also a little unusual in that the airfoils were designed specifically for the airplane. Generally, you'll find that when the engineer sits down to design an airplane, he says "What's the best airfoil?" and he looks at what's available and everything's a compromise. One's better for climb, one's better for cruise, one's better for stall and so on. Then he makes a selection from existing airfoils.

Well, that's what I originally did for the Defiant. The airfoil I picked for the back wing was developed by Dr. Eppler here before we built the wings for this airplane, and I complimented him on the fact that, in my data search, his airfoils were best for the job. Instead of smiling back at me and saying "Thank you," he said, "No, no! You shouldn't *pick* an airfoil for an airplane, you should *design* an airfoil for an airplane. What are your design requirements?" I said, "The thing I want is single-engine rate of climb. I want a Reynolds number of 1 million, 8/10 lift coefficient and max L over D at that point. Everything else is a compromise."

So he went back home to Germany, ran his computer program, and developed three brand-new airfoils optimized for that airplane. Not only for the airplane, but for the single-engine climb condition. He sent me the data and the airfoils and said, "This is it, this is better." Sure enough, I compared it to what I had, and it *was* better.

A lot of people are disbelievers. They don't think you can build a twin with a decent useful load and a decent range that will climb on one prop with the other one windmilling.

Actually, Rutan's intention was to make the Defiant as simple as possible, so that, if either engine failed, the plane would continue to fly without any corrective action on the part of the pilot.

The prototype and homebuilt models have fixed main gear and a retractable nose gear. The 160-hp carbureted Lycoming engines drive fixed-pitch props. There are no cowl flaps or retractable wing flaps. In short, there is not the usual series of controls the pilot has to fiddle with in other twins—quickly and in just the right sequence, if an engine fails at a critical point.

However, the production version of the Defiant was to have retractable main gear as well as retractable nose gear, and the power train was to consist of 180-hp engines and constant-speed props. Obviously, this added complexity would indeed require corrective action by the pilot if an engine failed. I asked Rutan about this at the time he was still hoping to market the production Defiant.

> In the past, there was no engine-out procedure. That's not true any more [of the production model]. There will be an engine-out procedure, but getting the gear up immediately in order to climb is not part of it, and the identification of the failed engine is something that can be left until some point in time later.

The Defiant can be thought of as having a cleanup procedure that doesn't have to be done immediately, or in the right order, to survive at low altitude.

Will we see a factory-made composite twin of canard design? We will, if Beech produces the Starship as expected. But the Starship is not an economical four-place plane with centerline thrust. For now, if that's what you want, you'll have to build it yourself.

The Lear Fan. This was the final product of the imaginative and resourceful Bill Lear, who died in March 1977, four years before the plane flew.

The Lear Fan is made of graphite/epoxy composites said to be four times stronger by weight than aluminum (Fig. 15.2). Because of its extreme rigidity, however, some questions have been raised about the material's crashworthiness, since it will not crumple and absorb energy in the same manner as aluminum.

If you were to examine this twin-engine airplane, you might find yourself wondering, "Where's the other prop?" There isn't any other prop. The Lear Fan is powered by two Pratt & Whitney PT-6 engines, each capable of 850-shp but flat-rated to deliver only 650-shp. The engines, located in the aft end of the fuselage, are connected to a transmission that drives a single four-blade propeller.

The Y-tail is also sort of different. (See Chapter 3.)

The plane has a maximum cruise of 363 kn at 20,000 ft. With its pressure differential of 8.6 psi, a sea-level cabin can be maintained up to 22,500 ft, and at its maximum certified altitude of 41,000 ft, the cabin altitude is only 8000 ft.

One of the Lear Fan's strong points is said to be fuel efficiency. The manufacturer points out that, on a 350-mi flight, the plane will

15.2. The Lear Fan. (*Lear Fan, Ltd.*)

take 13 min longer to reach its destination than a Learjet 25, but it will burn 200 gal less fuel. (The whole concept is a bit staggering to this pilot, who thinks more in terms of about 30 gal of fuel, *total*, for a 350-mi flight.)

At this writing, there is some question as to when, if ever, the Lear Fan will be produced. The bulk of the financing was raised by the British government, with the expectation that the plane would be manufactured in Northern Ireland. However, there have been certification difficulties in this country, accompanied by massive cost overruns. The pillar of strength that has kept the project alive in the face of many setbacks is a grey-haired lady named Moya Lear, who is determined to see Bill's last dream fly.

Starship 1. On August 29, 1983, what looked like the biggest homebuilt ever emerged from a massive hangar next door to the Rutan Aircraft Factory in Mojave, California, and made its first flight.

It had the trademarks of Burt Rutan, guru of the do-it-yourself planemakers: composite construction, a canard up front, a large swept main wing with winglets aft, and pusher propulsion.

But this was no basement project for a Bunyonesque homebuilder, destined to be shown off at the next EAA convention. Its mission was to provide posh transportation to corporate VIPs, in the style befitting a $2.7 million airborne limousine.

It was the Beech Starship 1 (Fig. 15.3).

More accurately, it was an 85 percent scale prototype, built by Burt Rutan's *other* company, Scaled Composites, Inc., for Beech Air-

AIRPLANES OF A DIFFERENT SORT

craft. Did Rutan actually design the Starship? That's a question I've never had answered to my complete satisfaction. Shortly after the scale prototype flew, I asked Rutan, but he had an agreement with Beech to let them do the talking. So I flew to Wichita and asked Beech's then-president Linden Blue. I got a very lengthy answer, which boiled down to the premise that some time in the late 1970s Beech assigned its engineers the task of developing a state-of-the-art plane, and they came up with a number of ideas, including the canard and aft engines. Then they called in Rutan, since he had done a lot of work in this area, and he contributed a lot to the development of the design, in conjunction with Beech's people.

Anyway, the full-scale Starship 1 will be built of carbon fiber, Kevlar, fiberglass, and titanium. The main wing will have a span of 54 ft 7 in., with 7 ft 9 in. tipsails (giant winglets with rudders). The canard will be a variable geometry design; that is, it will be in a forward position during takeoff and landing for maximum lift, and swept back in flight for reduced drag.

15.3. 85%-scale prototype of the Beechcraft Starship 1. (*Beech Aircraft Corp.*)

Power will come from two Pratt & Whitney PT6A-60 turbine engines, flat-rated to 1000 shp each. The engines will be in pusher configuration, mounted inboard and above the trailing edge of the aft wing.

The aircraft will have an 8.5 psi cabin-pressure differential, resulting in an 8000-ft cabin altitude at 41,000 ft. Rate of climb is expected to be 3300 fpm, with a maximum cruise of more than 348 kn.

Gross weight will be 12,500 lb. The cabin interior will be nearly a foot wider and 2 ft longer than that of Beech's current top-of-the-line Super King Air B200 turboprop. A typical passenger configuration will include a four-seat forward club section, with seating for three more on an aft seat that faces a two-place divan.

Up front, there will be a "glass cockpit," with the new EFIS (electronic flight instrument system) with cathode ray tube displays for navigation and performance monitoring systems.

Oh, yes. I also asked Linden Blue if Beech plans to make composite light planes. His reply:

First of all, let's understand something about graphite: it's very expensive. In terms of raw materials, the cost of graphite is five to ten times that of aluminum. For a high-performance airplane like the Starship, where weight is so important, it pays to have graphite. But we're talking about airplanes in the $2–$3 million class.

Somewhere down the road, graphite airplanes will be less labor intensive. Right now, they're *more* labor intensive. We hope to do enough learning on this, and enough refinement of production techniques, and enough mass buying of materials that we would ultimately get the price down to where we can make smaller airplanes. But it's not here yet.

EPILOGUE

A FEW FINAL THOUGHTS:

In the Introduction, I expressed the opinion that operating handbooks don't tell you as much as you ought to know about the equipment. That doesn't mean you should ignore the handbooks.

A fellow I know at Cessna conducts ground-school classes the factory offers to owners and pilots of Cessna's more sophisticated aircraft. From time to time he has asked a class for a show of hands by those who have not read the pilot's operating handbook for the plane they've been flying for maybe several hundred hours. He sees a lot of hands.

That's not good. If you do not read the manuals, you are definitely missing plenty, not only from a safety standpoint, but for getting the most value out of that equipment. This applies not just to the airplane and engine but, very important, to the avionics and instruments.

If the literature still leaves you unsatisfied, and you can't get the answers from your dealer, call the manufacturer. They have customer reps who are there to help.

And read before you buy. When you're evaluating a piece of equipment, be objective. Make a checklist of your real wants and needs for today and tomorrow, and see if what you're looking at really fills the bill. Then look at some comparable equipment and reevaluate.

It's a little like going to ground school all over again. But it will pay off—every time you fly.

DIRECTORY OF MANUFACTURERS

Advanced Aircraft Corporation
2016 Palomar Airport Road
Carlsbad, California 92008
619-438-1764

Aerotrim, Inc.
1130 102nd Street
Bay Harbor Island, Florida 33154
305-864-3336

Aire-Sciences, Inc.
216 Passaic Avenue
Fairfield, New Jersey 07006
201-228-1880

AiResearch Industrial Division
Garrett Corporation
3201 Lomita Boulevard
Torrance, California 90505
213-530-1981

Alcor, Inc.
10130 Jones-Maltsberger
San Antonio, Texas 78284
512-349-3771

American Navion Society
PO Box 1175
Banning, California 92220
714-849-2213

Ameromod Corporation
Building C64, Paine Field
Everett, Washington 98204
206-353-3559

ARNAV Systems, Inc.
PO Box 23939
Portland, Oregon 97223
503-684-1600

Astronautics Corp. of America, Inc.
PO Box 523
Milwaukee, Wisconsin 53201
414-671-5500

Avco Lycoming
Williamsport Division
652 Oliver Street
Williamsport, Pennsylvania 17701
717-323-6181

Avcon/Bush
PO Box 654
Udall, Kansas 64146
316-782-3317

Aviation Enterprises
2870 East Wardlow
Long Beach, California 90807
213-429-5949

Avionics West
3233 Skyway Drive
Santa Maria, California 93455
805-928-3601

Beech Aircraft Corp.
Wichita, Kansas 67201
316-681-7111

Bellanca Aircraft Corp.
Alexandria, Minnesota 56308
612-762-1501

Bendix Corp.
Avionics Division
PO Box 9414
Fort Lauderdale, Florida 33310
305-776-4100

Beryl D'Shannon
8220 220th Street West
Airlake Industrial Airport
Lakeville, Minnesota 55044
612-469-4783 / 800-328-4629

Brittain Industries, Inc.
PO Box 51370
Tulsa, Oklahoma 74151
918-836-7701

DIRECTORY OF MANUFACTURERS

C. S. Industries
7420 A South 199th Street West
Viola, Kansas 67149
316-545-7158

Cameron Aircraft Interiors
Du Page Airport
West Chicago, Illinois 60185
312-584-9359

Century Aircraft
10001 American Drive
Amarillo, Texas 79111

Century Flight Systems
PO Box 610
Municipal Airport
Mineral Wells, Texas 76067
817-325-2517

Cessna Aircraft Comp.
PO Box 1521
Wichita, Kansas 67201
316-685-9111

Colemill Enterprises
PO Box 60627
Cornelia Fort Airpark
Nashville, Tennessee 37206
615-226-4256

Collins General Aviation Division
Rockwell International
Cedar Rapids, Iowa 52406
319-395-1000

Composite Aircraft Corp.
523 Ridgeview Drive
Florence, Kentucky 41042
606-371-7247

Continental
    See Teledyne Continental Motors

Controlled Flight Mechanisms
PO Box 135
Galesville, Maryland 20765
301-867-1619

Custom Aircraft Conversions
254 West Turbo Drive
San Antonio, Texas 78216
512-349-6358

David Clark Co., Inc.
360 Franklin Street
Worcester, Massachusetts 01604
617-756-6216

Davtron, Inc.
427 Hillcrest Way
Redwood City, California 94062
415-369-1188

Dyna-Cam
PO Box 12095
Santa Ana, California 92712

EAA Aviation Center
Wittman Airport
Oshkosh, Wisconsin 54903-2591
414-426-4800
(Direct Autogas inquiries to:
Auto Fuel Research Department)

Electronics International, Inc.
5289 NE Elam Young Parkway #G200
Hillsboro, Oregon 97123
503-640-9797

Bob Fields Aerocessories
PO Box 390
Santa Paula, California 93060
805-525-6236

Flight Bonus
PO Box 120773
Arlington, Texas 76012

Flint Aero
8665 Mission Gorge Road
Building D, Unit 1
Santee, California 92071
619-448-1551

Foster AirData Systems, Inc.
7020 Huntley Road
Columbus, Ohio 43229
614-888-9502

Fredrickson Communications, Inc.
11100 West 82nd Street, Suite 104
Lenexa, Kansas 66214
913-492-6388

B. F. Goodrich
500 South Main Street
Akron, Ohio 44318
216-374-3600

DIRECTORY OF MANUFACTURERS

(Grumman American)
Gulfstream American Corp.
PO Box 2206
Savannah, Georgia 31402
912-964-3000

Hartzell Propeller, Inc.
PO Box 919
Piqua, Ohio 45356
513-773-7413

Horton, Inc.
Wellington Municipal Airport
Wellington, Kansas 67152
316-326-2241

Insight Instrument Corp.
PO Box 194, Ellicott Station
Buffalo, New York 14205-0194
416-871-0733

Isham Aircraft
Midcontinent Airport
PO Box 12172
Wichita, Kansas 67277
316-755-0713

King Radio Corp.
400 North Rogers Road
Olathe, Kansas 66061
913-782-0400

KNOTS 2 U, Inc.
1941 Highland Avenue
Wilmette, Illinois 60091
312-342-6550

Lake Aero Styling & Repair
PO Box 545
Lakeport, California 95453
707-263-0412

Laminar Flow Systems
PO Box 8557
St. Thomas, Virgin Islands 00801
809-775-5515

Lear Fan, Ltd.
PO Box 60000
Reno, Nevada 89506
702-972-2787

Lompoc Aero Specialties
Lompoc Airport
PO Box 998
Lompoc, California 93436
805-736-1273

Lycoming
  See Avco-Lycoming

McCauley Accessory Division
Cessna Aircraft Company
1840 Howell Avenue
Dayton, Ohio 45417

Machen Industries
3608 Davison Boulevard South
Spokane, Washington 99204
206-838-5326

Marsh Aviation
5060 East Falcon Drive
Mesa, Arizona 85205
602-832-3770

Maule Aircraft Corp.
Spence Air Base
Moultrie, Georgia 31768
912-985-2045

Met-Co-Aire
PO Box 2216
Fullerton, California 92633
714-870-4610

Micrologic
20801 Dearborn Street
Chatsworth, California 91311
818-998-1216

Miller Air Sports
San Marcos Municipal Airport
Route 2, Box 356 D
San Marcos, Texas 78666
512-353-7422

J. W. Miller Aviation
PO Box 7757
Marble Falls, Texas 78654
512-598-2556

Mooney Aircraft Corp.
PO Box 72
Kerrville, Texas 78028
512-896-6000

Narco Avionics
Fort Washington, Pennsylvania 19034
215-643-2900

Nelco
7095 Milford Industrial Road
Baltimore, Maryland 21208-6094
301-484-3284

Ted Nelson Co.
PO Box 20637
Reno, Nevada 89510
702-323-4955

Offshore Navigation, Inc.
PO Box 23504
Harahan, Louisiana 70183
504-733-6790

Petersen Aviation
Route 1, Box 18
Minden, Nebraska 68959
308-832-2200

Piper Aircraft Corp.
PO Box 1328
Vero Beach, Florida 32961
305-567-4361

Plantronics
345 Encinal Street
Santa Cruz, California 95060
408-426-5858

Precise Flight
63120 Powell Butte Road
Bend, Oregon 97701
503-382-8684 / 800-547-2558

Puritan-Bennett Aero Systems Co.
111 Penn Street
El Segundo, California 90245
213-772-1421

Radio Systems Technology
13281 Grass Valley Avenue
Grass Valley, California 95945
916-272-2203

RAM Aircraft Modifications
Waco-Madison Cooper Airport
PO Box 5219
Waco, Texas 76708
817-752-8381

Ramshead Exclusives
3070 Kerner Boulevard
San Rafael, California 94901
415-457-2772

Revere Electronics, Inc.
24118 Woodway
Cleveland, Ohio 44122
216-382-8819

Roto-Master
7101 Fair Avenue
North Hollywood, California 91605
818-982-4500

Rutan Aircraft Co.
Bldg. 13, Airport
Mojave, California 93501

R/STOL Systems, Inc.
Snohomish County Airport
North Complex C-72
Everett, Washington 98204
206-355-0645

Safe Flight Instrument Corp.
New King Street
White Plains, New York 10620
914-946-9500

Scott Aviation Products
225 Erie Street
Lancaster, New York 14086
716-683-5100

SDI-Hoskins
1762 McGaw Avenue
Irvine, California 92714
714-546-0601

Seguin Aviation
2075 Highway 46
Seguin, Texas 78155
512-379-3278

Sensenich Corp.
Airport Road
PO Box 4187
Lancaster, Pennsylvania 17604
717-569-0435

Shadin Company, Inc.
6950 Wayzata Boulevard, Suite 206
Minneapolis, Minnesota 55426
612-544-6422 / 800-328-0584

DIRECTORY OF MANUFACTURERS

Sigtronics
824-A Dodsworth Avenue
Covina, California 91724
818-967-0977

Silver Instruments, Inc.
Oakland Airport Industrial Park
8208 Capwell Drive
Oakland, California 94621
415-638-5600

Sky Ox, Ltd.
PO Box 500
St. Joseph, Michigan 49085
616-925-8931

Smith Speed Conversions
PO Box 430
Johnson, Kansas 67855
316-492-6254

Soloy Conversions
PO Box 60
Chehalis, Washington 98532

Sperry
5355 West Bell Road
Glendale, Arizona 85308
602-863-8000

S-TEC Corp.
Wolters Industrial Complex
Route 3, Building 946
Mineral Wells, Texas 76067
817-325-9406

Superplane, Inc.
22962 Clawiter Road #24
Hayward, California 94545
415-782-5425

Taylorcraft Aviation Corp.
1460 Commerce NE
PO Box 243
Alliance, Ohio 44601
216-823-6675

Teledyne Continental Motors
Aircraft Products Division
PO Box 90
Mobile, Alabama 36601
205-438-3411

Telex
9600 Aldrich Avenue South
Minneapolis, Minnesota 55420
612-884-4051

Terra Corp.
3520 Pan American Freeway
Albuquerque, New Mexico 87107
505-884-2321

Texas Instruments, Inc.
PO Box 405, MS3438
Lewisville, Texas 75067
214-462-5220

3M Stormscope
4800 Evanswood Drive
Columbus, Ohio 43229
614-885-3310

Thuuder Engine Corp.
7120 Hayvenhurst, Suite 321
Vau Nuys, California 91406

TKS, Ltd.
Kohlman Aviation Corporation
319 Perry Street
Lawrence, Kansas 66044
913-843-4098

Turboplus
Tacoma Narrows Airport
1520 26th Avenue
Gig Harbor, Washington 98335
206-851-6440

II Morrow, Inc.
PO Box 13549
Salem, Oregon 97309
800-742-0077 / 503-581-8101

Univair
Route 3, Box 59
Aurora, Colorado 80111
303-364-7661

White Diamond Corp.
PO Box 646
Chatsworth, California 91311
818-883-7191

Windecker Eagle
— *See* Composite Aircraft Corporation

Wings With Springs
PO Box 495
R.D. 3
Pleasant Unity, Pennsylvania 15676
412-423-2249

Wulfsberg Electronics
11300 West 89th Street
Overland Park, Kansas 66214
913-492-3000

INDEX

A&P (airframe and powerplant) mechanic, 25, 36
ADF (automatic direction finder), 25, 101–2, 142, 161
Advanced Aircraft Corp., turboprop modifications, 160
Advanced design aircraft
 canard design, 175, 178
 composites, pros and cons, 171–72, 176–78
 examples, 170–80
 unconventional engine configurations, 175, 178, 180
Aero Commander, 37
Aerostar, 44, 51, 63, 165
Aero Tech publications, 29
Aero-Trim, Inc., aileron trim systems, 160
Age-Master, 89
Ag-planes (agricultural), 43, 56
Ailerons, 9, 13, 16–18
 control force reduction, 17
 drooped aileron modification, 166
 Frise, 16, 17, 18
 interconnect with rudder, 27
 snatching, 17
 types, 16
Aircraft Electronics Association (AEA), 127, 129
Aircraft Owners and Pilots Association (AOPA), 28, 130, 142
 AOPA Pilot Magazine, 169
Aire-Sciences, Inc., avionics RT 553A/ RT 563A, 100
AiResearch, turbocharging systems, 60, 64, 162
Airflow separation, 4, 8, 10. *See also* Laminar flow; Turbulent flow
Airfoils
 angle of attack, 4
 design by computer, 10, 11, 12, 176
 development, 4–11, 176
 four-digit, 6
 how they work, 3–4
 laminar flow. *See* Laminar flow
 measurement by Reynolds Number, 6, 8–9
 numbering system, 6
 six-digit, 6
 turbulent flow. *See* Turbulent flow
 wing cross sections, 6–7
Airframe and powerplant mechanic. *See* A&P
Airspeed indicator. *See* ASI
Air traffic control. *See* ATC
Airworthiness Directives (ADs)
 elimination through prop mod, 161
 importance in purchasing aircraft, 28
Alcor, Inc., 30, 32, 33, 34
 Combustion Analyzer, 134
 exhaust gas temperature (EGT) gauge, 133
 Tru-Flow, 140
Allison aircraft engines, 165, 168, 173
Alther, George, 171
American Navion Society, modifications, 160
American Society for Testing Materials. *See* ASTM
Ameromod Corp., modifications, 161
Anderson Greenwood Aries T-250, 20
Angle of attack, 4, 8, 26
 indicator instruments, 142
Antennas
 installed in composite aircraft, 171
 radar, 111
Anti-icing and de-icing equipment, 85
 icing conditions, 85–86
 maintenance, 89
 new developments, 90–91
 operating techniques, 87
 types of ice, clear, rime, 88
 types of systems, 86–87
AOPA Pilot magazine, 169
ARC (Aircraft Radio Corporation), 94, 100, 128. *See also* Sperry
Area Navigation Systems (RNAV), 92, 101, 103–6
ARNAV Systems, Inc.

ARNAV Systems, Inc. (*cont.*)
  fuel flow meters, 140
  R-21, 108
  R-21/NMS, 93, 108
  R-40, 108
ASI (airspeed indicator), 15, 141–42
ASTM (American Society for Testing Materials), 40
Astronautics Corp. of America, Inc., 121, 150
ATC (air traffic control), 31, 88, 103
ATIS (automatic terminal information service), 96
Attenuation, 110, 121, 122, 123, 126
Autogas. *See* Fuel, aviation
Automatic direction finder. *See* ADF
Automatic terminal information service. *See* ATIS
Autopilot
  flight director, 129, 144, 148, 149, 150, 153–55
    double-cue, 153–54
    pros and cons, 154–55
    single-cue, 153–54
  how to choose, 150
  operating tips, 151
  reasons to have, 144
  types of systems, 100, 144–50. *See also* Century Flight Systems, Century models; King/Bendix; S-TEC Corp.
    position-based, 145
    rate-based, 145, 148
    two-axis vs three-axis, 145
Avco Lycoming. *See* Lycoming
Avcon/Bush, modifications, 161
Avgas. *See* Fuel, aviation
Aviation Consumer, The, 23, 28
Aviation Enterprises, modifications, 161
Avionics
  airborne telephones, 119–21
  area navigation (RNAV), 103–6
  automatic direction finder (ADF), 25, 101–2, 142, 161
  cigarette smoke, problems, 132, 151
  comms and navs, 95–100
  cooling, importance, 131
  digital displays, 93
  distance measuring equipment (DME), 95, 101, 103, 104, 105, 131
  entertainment systems, 125–26
  headsets, 121–22
  installation, 126–31
    factory vs field, 126–30
    how to find a good shop, 129
    interconnect drawing, importance, 129
    warranty considerations, 127, 129, 130
  intercoms, 123–25
  Loran, 106–9
  maintenance tips, 126, 131–32
  master switch, importance, 132
  microphones, 122–23
  microprocessor technology benefits, ix, 92–93, 137
  navs and comms, 95–100
  prices, ix
  purchasing tips, 130–31
  radar, 109–16
  radar antennas, 111
  Stormscope, 116–19
  transponders, 102–3
  weather radar, 109–16
Avionics West, entertainment systems, EC-200, 125

Beech Aircraft Corp., 19, 22, 24, 30, 87, 125, 127–28, 157
  Baron, 9, 16, 17, 33, 43, 33, 50, 78, 87, 91, 128, 161, 162, 168
  Bonanza, 9, 14, 15, 16, 17, 18, 27, 33, 39, 43, 44, 50, 58, 87, 91, 92, 128, 158, 160, 161, 162, 164, 165, 166, 168, 170
    V-Tails, 22–24, 161
  custom radio installations, 127–28
  Debonair, 22
  Duchess, 10, 16, 20, 50, 87, 91
  Duke, 9, 50, 87, 128, 161
  King Air, 9, 14, 20, 50, 87, 128, 180
  Lightning, 78, 173
  Sierra, 10, 14, 16, 26, 50, 52
  Skipper, 11. 14, 20, 52
  Starship, 25, 172, 177, 178–79
  Sundowner, 10, 14, 26, 30, 52
  Travel Air, 161, 168
Bellanca, August, 172, 173, 174
Bellanca, Giuseppe, 173
Bellanca Aircraft Corp., 20
  Cruisair, 52, 173
  Cruisemaster, 173
  Junior, 173
  Skyrocket, 174
  Viking, 78, 173

INDEX

Bendix Corp.
   Aerospace Division, 94
   Avionics Division. *See* King/Bendix
RDS-82 radar, 113
Bernoulli, Daniel, 3, 4
Blue, Linden, 179, 180
Boeing, 757/767, 22
Boyle's Law, 79
Brittain Industries, Inc., 150, 160

Cabin pressurization. *See* Pressurization
Camber, 4, 6, 9, 10
Cameron Aircraft Interiors, 162
Canard, 25, 26, 175
   disadvantages, 26
   stall prevention characteristics, 26
Carburetor systems
   carburetor ice, 30
   conversions to fuel injected, 162
   vs fuel injection systems, 30–31
   fuel jet constrictions, 136
   mixture control, 31–33
Casa, 46
Cathode ray tube. *See* CRT
CDI (course deviation indicator), 95, 97, 98, 100, 103, 104, 105, 106, 107, 147, 151, 153, 154
Center of gravity. *See* CG
Century Aircraft, turbocharging system installations, 63
Century Flight Systems autopilots, 151, 154
   Century models I, II, III, IV; 146
   Century models 21, 31, 41; 146
Cessna Aircraft Company, 43, 50, 75, 77, 86, 100, 125, 127, 128, 142, 158, 163, 165, 175, 181
   Airmaster, 6
   Citation I, 9, 20, 22
   Citation II, 9, 20, 22, 91
   Citation III, 20
   Crusader, 14
   custom radio installations, 128
   Hawk XP, 164
   Singles, 15, 52, 162, 163, 165, 166
   Skymaster, 81, 83, 141, 163, 175
   140, 56
   150, 9, 39, 56, 162
   152, 9, 18, 56, 161, 162
   170, 56, 161, 169
   172 (Skyhawk), 6, 9, 18, 50, 53, 56, 141, 144, 161, 162, 166, 168
   175 (Skylark), 169
   177 (Cardinal), 37
   180, 9, 56, 166, 169
   182 (Skylane), 6, 9, 18, 43, 50, 56, 60, 63, 86, 87, 91, 111, 162, 166, 169
   185, 18, 56, 162, 166, 168
   200 series, 160
   206, 9, 18, 87, 165, 168, 169
   207, 18, 165
   208 (Caravan), 18, 19
   210 (Centurion), 18, 50, 76, 77, 80–83, 87, 91, 160, 165
   300 series, 9, 14, 18, 22, 51, 53, 91, 162, 166, 168
   400 series, 9, 14, 18, 50, 166
CG (center of gravity), 55, 56, 67, 141
Champion Aircraft Corp., 52, 56
   Citabria, 56
   Decathlon, 56
   Scout, 56
Chord, 3, 4, 9, 11, 19, 174
CHT (cylinder head temperature) gauge, 32–33, 35, 134, 137
Civil Aeromedical Institute, address, 75
Clark, David, headsets and intercoms, ISOCOM, 123–24
Colemill Enterprises, engine and prop modifications, 162
Collins, General Aviation Div., Rockwell International, avionics, 93, 94, 128
   ANS-351, 106
   DCE-400, 95
   DME-451, 101
   IND-451,106
   VHF-251, 98
   VHF-253, 98–99
   VIR-351, 98, 99, 106
Comms (communication equipment), 95–100, 121, 123, 125, 131
Composite Aircraft Corp., 171
Composites. *See* Advanced design aircraft
Computers, use of in aircraft design, 11, 12, 176
Continental (Teledyne-Continental Motors), aircraft engines, 28, 29, 30, 33, 35, 36, 169, 170, 173
Controlled Flight Mechanisms, Lift Reserve Indicator, 142
Control surfaces, design, 10, 11, 12, 19. *See also* Ailerons, Flaps, Spoilers, Tails

Course deviation indicator. *See* CDI
Crosswind landings, 54-57
  landing gear, 56
CRT (cathode ray tube), 119, 180
CS Industries, engine modifications, 162
Curtiss-Wright, aircraft engines, 37
Custom Aircraft Conversions, modifications, 162
Cylinder head temperature gauge. *See* CHT gauge

David Clark, headsets and intercoms, 123-24
Davtron, Inc., Instruments
  301 C, F, 143
  655-2, 143
  701 B, 142
  800, 143
  811 B, 143
  903, ID, 142
Deere Corp., wankel engines, 37
Defiant. *See* Rutan Aircraft Factory
Density altitude, 61, 62, 141, 143
Descent control
  by flaps, 15
  by spoilers, 18
Design by computer, 10, 11, 12, 176
Detonation, 33, 39
DG (directional gyro), 147, 148, 151, 152
Diamond, Bernard, M.D., 70
Dietrick, Jerry, 170, 172, 173
Directional gyro. *See* DG
Directory of manufacturers, 183-88
DME (distance measuring equipment), 95, 101, 103, 104, 105, 131
Downwash, 4, 20
Drag
  aileron, 17
  flap, 15
  friction, 8
  induced, 165, 175
  parasite, 165, 175
  propeller, 42
  reduction by composites, 171
  reduction by modifiers, 156, 162, 168, 175
  wing, 6, 8-10, 11
D'Shannon, Beryl, modifications, 161
Dyna-Cam. *See* Engines

EAA (Experimental Aircraft Association), 38, 39, 40, 130, 170

Eagle (Windecker), 170-73
E-A-R, ear plugs, 122
EAROM (electrically alterable read only memory), 96, 102
ECDI (electronic course deviation indicator), 97, 99, 100
Edo-Aire. *See* Aire-Sciences; Century Flight Systems, Century models
EFIS (electronic flight instrument system), 115, 180
EGT (exhaust gas temperature) gauge. *See* instruments
Electronics International, Inc., 33
  EGT/CHT, 137
Ellis, Dave, 14
Engines. *See also* Continental; Dyna-Cam; Franklin; Lycoming; Thunder; Wankel
  Airworthiness Directives (ADs), 28
  carburetion systems, 30
  configurations, unconventional, 25, 175, 178, 180
  cylinders, four vs six, 29
  design advances, 37-38
  diesel, 37
  Dyna-Cam, 37
  extending life, 29, 33, 35, 64
  failures on takeoff, 31
  fuel injected hot starts, 35
  geared prop, 31
  mixture control, 31-35, 133
  model numbers, interpretation, 30-31
  normally aspirated, 33, 62
  normally aspirated vs turbocharged, 58, 60-62
  oil, 29, 35-36, 64, 65
  power settings, 31
  preignition, 33, 36
  push/pull and pusher configurations, 25, 175, 178, 180
  rust inhibition, 36
  Service Difficulty Report (SDR), 28
  temperature, importance, 31-32
  thermal shock avoidance, 31, 53
  throttle technique, 31, 62, 63, 64
  time between overhauls (TBOs), viii, 29, 65
  turbocharged. *See* Turbocharging
  water-cooled, 36
  wear rate increase, 29
Entertainment systems, 123, 125-26
Eppler, Richard, 11, 176
Ercoupe, 53

INDEX

ETA (estimated time of arrival), 101, 119
ETE (estimated time enroute), 106, 107
Exhaust gas temperature (EGT) gauge. *See* Instruments
Experimental Aircraft Association. *See* EAA
Experimental Aircraft Association Aviation Federation, 38
EZ/Ox oxygen nasal cannula, 70

FARs (federal aviation regulations), 39, 64, 66, 70, 86
FBO (fixed base operator), 40, 74, 129
FCC (Federal Communications Commission), 120
Federal Aviation Administration (FAA)
  Airworthiness Directives (ADs), 28
  approval to modify aircraft, 157, 159
  autogas, policy on, 38–39
  cabin noise levels, concern with, 122
  certification, 170
  federal aviation regulations (FARs), 39, 64, 66, 70, 86
  GADO (General Aviation District Office), 159
  high altitude chamber training, 74–75
  plans regarding Loran, 107, 109
  Service Difficulty Reports (SDRs), 28
  supplemental oxygen use, rules on, 66, 67, 68, 69–70
Federal aviation regulations. *See* FARs
Federal Communications Commission. *See* FCC
Fields, Bob, Aerocessories, noise reducing door seals, 162
Fixed base operators. *See* FBOs
Fixed gear, 47–49
Flaps, 13–16
  extension speed, 15
  extension systems, 15
  Fowler, 11, 14, 166
  modification, 159, 166, 176
  plain, 14
  rearward action, 13, 14
  slotted, 14
  split, 14
  upward deflecting, 11
  use as dive brakes, 31
Flight bonus, modifications, 162–63
Flight Director. *See* Autopilot

Flint Aero, fuel tank kits, 163
Fokker, Tony, 4
Foster AirData Systems, Inc., navigation systems
  LNS 616, 109
  RNAV 511, 105
  RNAV 612, 106
Fowler flap, 11, 14, 166
Franklin aircraft engines, 28
Fredrickson Communications, Inc., airborne telephones, 121
Frise ailerons, 16, 17, 18
Fuel, aviation (avgas), 30, 38
  additives
    fillers, 40
    lead with chlorine, bromine, 39
    TCP (tricresyl phosphate), 30
  auto fuel (autogas or mogas), use of, pros and cons, 38–40
  availability, 30, 39
  carburetion systems, 30
  computers and flow meters, 137
  diesel engines, savings with, 37
  economy through modifications, 158, 177–78
  economy through power settings, 31–34
  filtering advisability, 40
  injection systems, 30
  jet constriction, 136
  lean misfire, 32
  methane, 30
  mixture control, 31–33
  rich misfire, 32
Fuel/air ratio, 32, 37
Fuel computers, 137–41
  cautions on use, 138–40
Fuel injection systems, 30
Fuel injector, 30
Fueltron. *See* Silver Instruments, Inc.

GADO. *See* General Aviation District Office
Garrett Corp., 60, 83
Gasoline. *See* Fuel, aviation
Gauges. *See* Instruments
Gear
  landing. *See* Landing gear
  tailwheel. *See* Tailwheel gear
  tricycle. *See* Landing gear
General Aviation District Office (GADO), 159
Geronimo. *See* Seguin Aviation

Goodrich, B. F., anti-icing and de-icing equipment, 89, 90
Graham's Law, 79
Graphic Engine Monitor (GEM), 134–36
Greene, Leonard M., 141
Greenwich mean time (GMT), 101, 143
Griswold, Jim, 17
Grumman American, 15, 52, 161

Harlamert, W. B., 44
Hartzell Propeller, Inc., 42, 169
  composite blade, 46
  Q-Tip, 44
Headsets, 121–22, 124, 125
"Hershey bar" rectangular wing, 19, 163, 164
Hoerner wingtips and tiptanks, 165, 168
Hoffman propeller, 44
Horizontal situation indicator. See HSI
Horton, Inc. STOL Craft, modifications, 163–64
HotProp, anti-icing system, 87, 90
Hot starts, 30, 35
HSI (horizontal situation indicator), 103, 104, 105, 115, 119, 129, 148, 150, 151–53
Hundere, Al, 32
Huntington, Morgan, 142
Husick, Chuck, 43
Hypoxia, 66, 67, 74, 79

Icex, de-icing product, 89
Icing conditions. See Anti-icing and de-icing equipment
IFR (Instrument Flight Rules), 24, 31, 107, 108, 109, 122, 131, 144, 150
ILS (Instrument Landing System), 154, 155
Insight Instrument Corp., Graphic Engine Monitor (GEM), 134–36
Instrument Flight Rules. See IFR
Instrument Landing System. See ILS
Instruments
  airspeed indicator (ASI), 141–42
  automatic direction finder (ADF), 25, 101–2, 142, 161
  clocks, 143
  course deviation indicator (CDI), 95, 97, 98, 100, 103, 104, 105, 106, 107, 147, 151, 153, 154
  cylinder head temperature (CHT) gauge, 32–33, 35, 134, 137

directional gyro (DG), 147, 148, 151, 152
electronic course deviation indicator (ECDI), 97, 99, 100
electronic flight instrument system (EFIS), 115, 180
exhaust gas temperature (EGT) gauge, 31–35, 133–37, 141
  multi-probe, advantages, 133
flight director. See Autopilot, flight director
fuel computers, 137
  cautions on use, 137–38, 140–41
fuel pressure gauge, 64
horizontal situation indicator (HSI), 103, 104, 105, 115, 119, 129, 148, 150, 151–53
outside air temperature gauge (OAT), 139, 142, 143
pressurization system indicators, 81–83
radio magnetic indicator (RMI), 100
stall indicators, 141–42
tachometer, 141
turbine inlet temperature (TIT) gauge, 33, 64, 141
turbocharged aircraft, 64
Insurance
  product liability, 37
  time-in-type requirements, 47
Intercoms, 123–25
Intercoolers, 64, 168, 169
Irwin, Jim, 144
Isham Aircraft, modifications, 164

John Deere Corp., 37

KIAS (knots indicated airspeed), 141
King/Bendix, avionics and autopilots, 93, 95, 111, 128, 131, 145, 146, 151
KA 20, 131
KAP 100/150/200, 148
KFC 150/200, 146, 148
KGR 356, 113
KMA 24, 121
KN 53, 96
KN 62A, 101
KN 63, 101, 105
KNS 80, 92, 104, 105
KNS 81, 105
KR 86/87, 102
KT 79, 102
KT 96, 120
KWX 56, 113

INDEX

KX 155/165, 95, 96
KY 196/196-05/196-10, 96
KY 197/197-05/197-10, 96
King Radio Corp. *See* King/Bendix
Knots indicated airspeed. *See* KIAS
Knots 2 U, Inc., modifications, 164
Koppers Aeromatic propeller, 43

Lake Aero Styling & Repair, modifications, 164
Lake Amphibian, 42
Laminar flow, viii, 6–11
  description, 8
  examples, 10, 11
  stall characteristics, 10
Laminar Flow Systems, modifications, 164
Landing, 3, 11, 47, 48, 51, 52, 56
  crosswind, 20, 55, 56
Landing gear, 47–57
  automatic extension systems, 50, 51
  dive brakes, using for, 31, 53
  electric vs hydraulic systems, 49–52
  emergency extension systems, viii, 49–52
  fixed, viii, 47–49
  ground clearance, 52
  ground loop, 55, 56
  Johnson-bar, 51, 52
  mistaking gear handle for flap handle, 15–16, 161
  retractable, viii, 47, 48, 49
  shock absorbers, 52
  taildragger vs tricycle, 47, 53–57
  turning radius, 52
LCD (liquid crystal display), 93, 99, 108, 119, 137, 140, 143
Lear, Bill, 177
Lear, Moya, 178
Lear Fan, 25, 78, 90, 172, 177–78
Learjet, 10
LED (light-emitting diode), 93, 100, 107, 108, 125, 139
Lift
  coefficient, 6, 10, 14
  creation, 4, 8
  vs drag in aileron design, 14, 17
  vs drag in wing design, 6, 10, 11
  search for maximum, 9
Light-emitting diode. *See* LED
Linton-Smith, Trevor, 17
Liquid crystal display. *See* LCD
Localizer (LOC), 100, 104, 142, 145, 146, 148, 149, 150, 151

Lompoc Aero Specialties, modifications, 165
Long Range Navigation. *See* Loran
Lopresti, Roy, 18, 44, 157, 164
Loran, 92, 93, 106–9, 140
Luscombe, 56
Lycoming, aircraft engines, 28, 30, 33, 36, 161–62, 165, 166, 169, 174, 176

McCauley Accessory Division, Cessna Aircraft Co., propellers, 44
Machen Industries, modifications, 165
Manifold pressure. *See* MP
Manufacturers' addresses. *See* Directory of Manufacturers
Manufacturers' operations manuals, viii, 15, 32, 64, 147, 151, 181
Marsh Aviation, modifications, 165
Maule Aircraft Corp., 27
Met-Co-Aire, aircraft parts, 165
Methane, 30
Micrologic, Loran systems, ML-6500, 108
Microphones, 121, 122–23
Mid-continent gap, 107, 109
Miller, J. W., Aviation, modifications, 165
Miller Air Sports, modifications, 165
Mitsubishi MU-2, 10, 18, 44
Mixture control, 31–35, 133
Modifications, aircraft. *See* Mods
Mods
  cost justification, 158–59
  how to choose, 159
  mod shops, viii, 18, 63, 68, 156, 158, 160–69
  performance improvements, 156–57
  types, 160–69
  utility and cosmetic improvements, 157–58
Mogas. *See* Fuel, aviation
Mooney Aircraft Corp., 27, 44, 50, 52, 87, 111, 125, 145, 162, 164, 165, 173
  Mite, 52
  M-20 Series, 25, 166
  Mustang, 58, 77
  201, 17, 44, 51, 60, 87, 146, 157, 166
  231, 17, 44, 51, 60, 62, 87, 157, 166, 169, 173
  301, 11, 14, 18, 44, 77
MP (manifold pressure), 62, 63, 64, 83

NACA (National Advisory Committee for Aeronautics). See also NASA
wing designs, 4-6, 9, 10, 46, 174
Narco Avionics, 93, 94, 95, 128
ADF 841, 102
COM 810/811, 97
COM 824/825, 98
DME 190/195, 94
DME 890, 101
Escort I and II, 96-97
IDME 891, 101
LRN 820, 108
Mark 12D, 97
NAV 824/825, 98
NS 800/801, 106
RNAV 860, 101
NASA (National Aeronautics and Space Administration), 17. See also NACA
propeller designs, 43, 44, 46
wind tunnels, 5
wing designs, 10-11
Nasal cannula, 70
Navion, 43, 49, 160
Navs (navigation equipment), 95-100
NDBs (nondirectional beacons), 102
Nelco, Loran systems
AF 921, 108
AF 92IR, 109
Nelson, Ted Co., Oxygen Flow Meter, 69, 70
Newton, Sir Isaac, 3, 4
Nixon, John, 154-55
Nondirectional beacons. See NDBS

OAT (outside air temperature) gauge, 139, 142, 143
OBS (omni bearing selector), 97, 100, 150
OEM (original equipment manufacturer), 63
Offshore Navigation, Inc., Loran systems, ONI-7000, 109
Omni bearing selector (OBS), 97, 100, 150
Original equipment manufacturer (OEM), 63
Outside air temperature gauge. See OAT gauge
Oxygen, 62, 64, 66-74, 76, 79
flow rates, recommended, 68
hypoxia, 66, 67, 74, 79
maintaining the system, 68, 70
manufacturers, 71-73. See also

Nelson, Puritan-Bennett, Scott, Sky-Ox, White-Diamond
masks, 69-70
nasal cannula, 70
need for, 66
night flying, 67
systems, 67-69, 72

Petersen Aviation, autogas STCs, 40
P-factor, 27, 55
Pilot reports. See PIREPs
Piper, Howard "Pug," 174
Piper Aircraft Corp., 26, 87, 111, 125, 127, 157
Aerostar, 44, 51, 63, 165
Apache, 165, 166
Archer, 17, 144, 159
Arrow, 17, 20, 21, 43, 51, 60, 62, 164, 166, 169
Aztec, 51, 165, 166
Cherokee (Vero Beach line), 15, 17, 19, 26, 50, 51, 161, 163, 164, 165, 166, 175
Cherokee Six, 21
Cheyenne III, 20, 44
Colt, 169
Comanche, 22, 162, 164, 165, 166
Cub, 39, 56, 169
custom radio installations, 128
Dakota, 17, 166, 169
Lance, 20, 63, 164
Malibu, 9, 14, 77, 80, 87
Navajo, 51, 162
PA 28, 164
PA 32, 164
Pacer, 53, 56, 169
Papoose, 172
Saratoga, 17, 21, 48, 51, 63, 68, 87
Seminole, 14, 17, 20, 51, 61
Seneca, 14, 15, 17, 18, 19, 43, 51, 62, 164, 166, 169, 175
Six 300, 21
Super Cub, 56
Tomahawk, 11, 17, 52, 161
Tri-Pacer, 53, 56, 169
Warrior, 164, 166
PIREPs (pilot reports), 86, 88
Pitts, 56
Plane & Pilot magazine, 169
Plantronics, headsets
MS-50, 124
Starset, 124
Power settings, 31

INDEX

Pratt & Whitney, engines, 160, 177, 180
Precise Flight, modifications, 18, 165, 169
Preignition, 33, 36
Pressurization, 62
 altitude changes, physiological effects, 78-79
 instruments, 81-83
 pressure differential, 80
 reasons for, 76-77
 systems, 79-83
Product liability, 37
Propellers, 3, 41-46
 airfoil, 41
 blades, number of, pros and cons, viii, 43
 care, 44-46
 composite blades, 46
 constant speed, 42
 fixed pitch, 42, 46
 full feathering, 42
 ground clearance, 43, 44, 54, 162
 movable pitch, 42
 noise reduction, 41, 43, 44
 non-rigid, 46
 proplet, 44, 45
 Q-Tip, 44
 reversible, 42
 spinner, 45, 46
Propwash, 15, 20
PTT (push-to-talk), 122, 126
Puritan-Bennett Aero Systems Co., oxygen systems, 72

Q-Star, 37

Radar, 109-16. See *also* Stormscope
 antennas, 111
 attenuation, 110
 cautions on use, 109-10
 color interpretation, cautions, 110-11
 how it works, 109-11
 stabilization, 113, 118
 vs Stormscope, 117
Radio common carriers. See RCC
Radio frequency interference. See RFI
Radio magnetic indicator. See RMI
Radios. See Avionics, individual types of equipment
Radio Systems Technology, kit-built avionics, 124
 RST-571/572, 100

Rajay, turbocharging equipment, 60, 62, 63, 72
RAM Aircraft Modifications, 166
Ramshead Exclusives, seat covers, 166
R & D (research and development), 46, 170
Rate controller, 81
RCC (radio common carriers), 120, 121
Reid vapor pressure. See RVP
Republic Seabee, 42
Research and development. See R & D
Retractable gear, viii, 47, 48, 49
Revere Electronics, Inc., headsets and intercoms, HUSH-A-COM, 124
Reynolds numbers, 6, 8-9
RFI (radio frequency interference), 121, 131
RMI (radio magnetic indicator), 97, 100
RNAV, 92, 101, 103-6
Roberts, Ron, 33
Robertson. See R/STOL
Rockwell International, 94
 Aero Commander, 37
 112, 22, 63
 114, 22
 700, 22
Roll control
 by ailerons, 17, 19
 by spoilers, 18, 19
Roto-Master, turbocharging systems, 60, 63
R/STOL Systems, Inc., modifications, 18, 166
Rudders
 aileron interconnect, 27
 effectiveness reduction by flaps, 15
 effectiveness reduction by tail design, 20
 effectiveness reduction by tailwheel landing gear, 55
 on Mooneys, 25
 ruddervators, 22, 168
Rutan, Burt, 11, 25, 26, 174, 175, 178
Rutan Aircraft Factory
 Defiant, 11, 26, 174-75
 Long-EZ, 175
 VeriEze, 25, 175
RVP (Reid vapor pressure), 39
Ryan, Paul, 116, 117

Safe Flight Instrument Corp., flight control systems, SC-150, 142

Sailplanes, 18, 19, 92
Scaled Composites, Inc., research and development, 178
Schiff, Barry, 15
Scott Aviation Products, oxygen systems, 72-74
SDI/Hoskins, fuel computers, 139-40
  CFS 1000A/1001A/2000A/2001A, 139
SDR. *See* Service difficulty report
Seaplanes, 42
Seguin Aviation, Geronimo, 166
Sensenich Corp., propellers, 161
Sensitivity time control (STC), 111
Service Difficulty Report (SDR), 28
Shadin Company, Inc., Digiflo, 139
Short takeoff and landing. *See* STOL
Sigtronics, STEREOCOM, 124, 125
Silver, Brent, 23
Silver Instruments, Inc., Fuelgard, Fueltron, 108, 140
Sky Ox, Ltd., oxygen systems, 72
Skyrocket, 78, 173-74
Smith, Mike, 78
Smith, Ted, Aerostar. *See* Piper Aerostar
Smith Prop-Jet, 78, 173
Smith Speed Conversions, 168
Soloy conversions, 168
Somers, Dan, 6, 11, 19
Spark plugs, 36
Sperry, 100, 111, 119, 128, 150
  Primus 100, 112
  Primus 150, 111, 114
  RN-478A, 106
  RTA-476A, 106
  WeatherScout, 111, 112
Spin, 25
Spitfire, 19
Spoilers, 13, 18-19, 165, 166
  slot-lip, 19
Sport Aviation magazine, 39, 40
Stabilator, 20, 26, 164
Stabilization, 4, 25, 27, 148, 165
  of radar, 113, 118
Stabilizer, 3, 20, 22, 25, 26, 164, 168
Stall, 3, 25, 56
  canard design, 26
  indicators, 141-42
  vs lift in wing design, 4, 6, 8-11
  unequal vs symmetrical, 10, 11
STC. *See* Supplemental type certificate; *See also* Sensitivity time control
S-TEC Corp.

autopilots, 144, 145
systems 40, 50, 60; 149-50
Stinson, 56, 169
STOL (short takeoff and landing), viii, 156, 164, 166
Stormscope
  how it works, 116
  vs weather radar, 117
  WX-8, 119
  WX-10/10A, 117-19
  WX-11, 118, 119
  WX-12, 119
Superplane, Inc., modifications, 168
Supplemental Type Certificates (STCs), 39, 40
  for autopilots, 146, 150
  for modifications, 159, 160, 161, 162, 164, 168, 169, 173
  for mogas, 30, 39
  for Q-Tip propellers, 44
  for turbochargers, 63
Swift, 55, 162

Tail
  canard, 26, 175. *See also* Canard
  cruciform, 22
  design considerations, 20
  Mooney, 25
  stabilators, 26
  T-Tails, viii, 20-21
  V-Tails, 22-24
  Y-Tails, 25, 177
Taildragger, 53-57, 158, 162, 169
Tailwheel gear, 53
Takeoff, 3, 20, 41, 42, 54, 55, 62, 63
Taylorcraft Aviation Corp., 56
TBO (time between overhauls), viii, 29, 65
TCA (terminal control area), 144
TCP (tricresyl phosphate) fuel additive, 30
Technical standard order. *See* TSO
Teledyne-Continental Motors. *See* Continental
Telephones, airborne, 119-21
Telex, headsets and intercoms, 125
Terminal control area. *See* TCA
Terra Corp., Loran systems, TXN 960, 99, 121
Texas Instruments, Inc., Loran systems
  TI 91, 108
  TI 9100/9100A, 108
  TI 9200, 108
3M Stormscope. *See* Stormscope
Thunder, aircraft engines, 36

INDEX

Time between overhauls. *See* TBO
Time-to-station. *See* TTS
TIT (turbine inlet temperature) gauge, 33, 64, 141
TKS Ltd., anti-icing equipment, 90, 91
Trade-a-Plane publication, 169
Trammell, Archie, 112, 113
Transponders, 102–3
Tricycle gear. *See* Landing gear
Trim systems, 25, 27, 160
TSO (technical standard order), 107, 108, 109, 131
TTS (time-to-station), 95, 101, 104, 106
Turbine inlet temperature gauge. *See* TIT
Turbocharging, viii, 58–65
  advantages and disadvantages, 60–62, 64
  coking, 64
  fixed orifice, 62
  how it works, 58–60
  manual wastegate, 62, 67
  pressurized aircraft, 67, 79, 83
  throttle-linked wastegate, 63
  turbine inlet temperature (TIT) gauge, 33, 64, 141
  types of systems, 62–64
Turboplus, modifications, 169
Turbo-supercharged, 30
Turbulent flow, 6–8
  definition, 8
  examples, 6, 9, 10
II Morrow, Inc., Loran systems
  602, 107
  611, 108
  612, 108

UHF (ultra high frequency), 120
Univair, aircraft parts, 169

Vapor lock, 30, 35, 39, 64
Vapor pressure, 39
Vertical navigation system. *See* VNAV
VFR (visual flight rules), 95, 103, 109, 122, 125, 144
VHF (very high frequency), 95, 109, 125
VHF omnidirectional range. *See* VOR
Visual flight rules. *See* VFR
VLF (very low frequency), 128
VNAV (vertical navigation system), 108
VOR (VHF omnidirectional range), 93, 95, 97, 98, 99, 100, 103, 104, 108, 142, 146, 147, 148, 149, 150, 151, 153, 154, 164
VORTAC (combined VOR and TACAN facilities), 96, 101, 103, 104, 105, 106, 109, 113

Wankel aircraft engines, 37
Warranties
  avionics, factory installed, 127, 130
  avionics, field installed, 128, 129, 130
  engine, TBO effect, 29
  engine development, problems with, 37
Waypoint, 92, 93, 103, 104, 105, 106, 107, 108, 109, 113
Weather radar, 109–16
Weight
  penalties
    anti-icing equipment, 91
    electrical gear systems, 49
    liquid-cooled engines, 36
    propellers, 43, 46
    tail design, 20
    tricycle gear, 54
  reduction
    with composites, 176, 177
    with fiberglass engine parts, 37
    with lightweight oxygen systems, 73
    with modifications, 162–69
    with unconventional designs, 176
White, Sidney, M.D., 70
White-Diamond Corp., nasal cannula, 70
Windecker Eagle, 170
Wind tunnels, 4–5
Wing
  airfoil cross sections, 6, 7
  design, 3, 19
  "Hershey bar" rectangular wing, 19, 163, 164
  lift, 6, 10, 11
  planform, 19
  semi-tapered, 19
  spar, 3, 4
  with springs, modifications, 169
Wright brothers, 4
Wulfsberg Electronics, airborne telephone, 119, 121

Yaw, 17, 43, 148

Zimmer, Ed, 94